THE REVOLT OF THE ENGINEERS

THE REVOLT
OF THE ENGINEERS

SOCIAL RESPONSIBILITY
AND THE AMERICAN ENGINEERING PROFESSION

EDWIN T. LAYTON, JR.

THE PRESS OF CASE WESTERN RESERVE UNIVERSITY
CLEVELAND AND LONDON/1971

TABLE OF CONTENTS

PREFACE

This book had its origins in my reading, as a student, Thorstein Veblen's *The Engineers and the Price System*. I was surprised that so acute an observer as Veblen would portray the engineers as the predestined leaders of a social revolution in America. When I came to examine the engineering press for the period in which Veblen wrote, I found that in the first quarter of the twentieth century the engineering profession was convulsed by a group of reformers, termed "progressives," who attempted to unite their profession and to use it as an instrument of social reform. These progressives were peculiar in several respects. Their emphasis on scientific planning, their use of democratic rhetoric, and their insistence on the need for drastic reform of the business system gave them an affinity with the contemporary progressive movement in national, state, and local politics. But engineers differed from other progressives. Despite their rhetoric, engineers had little faith in democracy as a social remedy, even though they might favor a sort of Greek democracy within their profession; instead they expected society to be saved by a technical elite.

In many ways engineering progressivism appeared to provide an interesting connecting link between the progressive period and the conservative reaction that followed it. For example, the conservation program, a major concern and inspiration of the engineers, has been classified by historians as progressive. But Herbert Hoover's ideas of industrial consolidation and efficiency have been considered conservative, if not reactionary. Yet conservation and Hoover's program could both be seen as manifestations of engineering progressivism. So, too, could scientific management and, to some extent, the technocracy movement of the 1930s. Because it did not fit the usual historical typology, the engineers' social movement appeared to offer insights into American history.

Another reason for examining the engineering progressives was that they promised to illuminate American technological and scientific development; I hoped that such a study would show, in a particular case, how these two great forces interacted with each other and with society. The very existence of engineering as a mass profession was a consequence of a profound revolution in technology. This involved an uprooting of technology from its matrix in centuries-old craft traditions and its grafting onto science. The change from art to science in engineering had been foreshadowed for centuries, but the large-scale application of science to practical affairs was for the most part a product of the nineteenth century. It was this change that brought engineers to the forefront of technology. And while this revolution was by no means complete by 1900, its potentialities had been amply demonstrated. It is not surprising, therefore, that engineers found in science a self-image and a mission. As stewards of technology, they were the agents of social progress. They held that as a group they had assimilated the rationality and objectivity of their discipline. They were convinced that social problems might be solved by applying the same method that had proved so fruitful in material affairs—that is, by engineering society. This set of ideas, which I have termed the "ideology of engineering," explained many of the peculiarities of engineering progressivism. But the two were by no means the same. Progressives were relatively a small group among engineers, but the ideology of engineering appeared to have been adopted by the great majority of engineers, judging by the materials I found in the technical press.

To account for engineering progressivism I found it necessary to see this social movement in the larger context of the professional development of engineering in America. The ideology of engineering served not only to define the engineer, his social role, and his social responsibilities, but also to develop his professional consciousness. Despite the very widespread adoption of these ideas, however, the engineers failed to develop the unity they were struggling for. Unlike doctors and lawyers, engineers were, for the most part, the employees of large organizations, chiefly private corporations. Professional development conflicted with the interests of their employers. The split that developed was not simply between engineers and businessmen. Engineers were tied to business by many bonds; not the least of these was the fact that many of the most successful engineers had gone into managerial work. These men did not repudiate the ideology of engineering, but instead they sought to use it to reaffirm the engineer's loyalty to his employer. Thus, engineers developed two versions of professionalism, one which directed the engineer to independent action and a second which linked the engineer to the business system. The story of the professional development of American engineering from 1900 to 1940 is in large part the story of the clash between these two traditions.

Engineering progressivism was a stage in the professional development of engineering in America. It had its origins in the 1880s and 1890s; by 1940 it was almost extinct. Reaching a peak during and immediately after the First World War, engineering progressivism appeared for a time to have succeeded. It united the fragmented engineering profession in the Federated American Engineering Societies. Under the leadership of Herbert Hoover, this organization attempted to apply engineering knowledge and methods to some of the great questions of the day. But its studies of waste in industry and the twelve-hour day aroused bitter resentment on the part of employers. These studies helped bring about a conservative reaction that crippled engineering progressivism. By 1940, the engineers had lost almost all of their reform zeal. Engineers turned to other strategies of professional development. Ironically, the greatest social influence of engineering progressivism came when the movement was declining among engineers. Through Herbert Hoover, these ideas gained currency in the business community in the 1920s, and they found many echoes in the reform proposals of the depression years.

Although engineers contributed to American reform, the true significance of their movement lay within the group, not in the larger society. Engineers did not succeed in preventing the abuse of technology nor in directing this force to social purposes. They failed to reform either themselves or society. But their search for personal dignity, independence, and social responsibility in a corporate age is still of relevance. It is not only for engineers that these goals are important.

ACKNOWLEDGMENTS

In writing a book one incurs debts to many people, debts that can never be fully acknowledged, much less repaid. Teachers, colleagues, students, librarians, and archivists have been generous with their time, and without their advice and encouragement this book could not have been written. I owe more to my teachers than to any others. Two deserve special mention. George Mowry directed the thesis out of which this book grew. But more importantly, he taught me to be a historian. John Higham introduced me to intellectual and social history. He not only had a lasting influence on my views of history, but it was he who first drew my attention to the history of technology as an important field for investigation. A number of colleagues, past and present, assisted me in many ways. Merle Curti encouraged me as a fledgling historian. My colleagues at Case Western Reserve University have been helpful on numerous occasions. I would particularly like to thank Melvin Kranzberg, Robert Schofield, Robert C. Davis, and Eugene Uyeki. I owe a special debt to Eugene Ferguson for encouraging my research and for reading this manuscript and making numerous suggestions. I claim, of course, responsibility for whatever errors remain.

This book benefited greatly from reading the works of scholars with allied interests. In some cases, I was also able to discuss common problems with them. In addition to those named above, I would like to mention Daniel C. Calhoun, Monte Calvert, George C. Daniels, Samuel P. Hays, Sam Haber, Milton J. Nadworny, Carroll Pursell, Bruce Sinclair, and Kenneth Trombley. To the late Morris L. Cooke I owe a unique obligation. He provided me with a vast fund of information through a series of interviews and an extensive correspondence over several years, and he also gave me friendly counsel and guidance. Since he was one of the principal actors in the story I have to tell, his help was of immense value.

Without the help of librarians, historians could not function. I have been assisted, with unfailing courtesy and patience, by so many that I cannot possibly list them all by name. My indebtedness is compounded by my use of both the main libraries of universities and departmental libraries in engineering and science. But I would like to thank the many librarians who helped me at the various libraries I have been associated with: UCLA, Wisconsin, Ohio State, Purdue, and Case Western Re serve University. In addition, I received courteous assistance at the Bancroft Library of the University of California, the libraries of Stanford and the University of Illinois, and the public libraries of Los Angeles, Cleveland, and New York.

A number of people have helped me gain access to manuscript materials. Morris L. Cooke, Mrs. Barbara Givan, B. F. Jakobsen, and Julian W. King very generously allowed me to use personal manuscript collections. I would like to thank Francis I. Duck and Samuel C. Williams at the Stevens Institute of Technology for assistance in using the Frederick W. Taylor papers and for helping to make my stay in Hoboken a pleasant one. The presidential libraries of the National Archives Service are in a class by themselves; in addition to the usual services, they helped me find accommodations and assisted me in other ways. In particular I would like to thank Mr. Dwight M. Miller of the Hoover Presidential Library for assistance in using the Herbert Hoover papers and Herman Kahn of the Franklin D. Roosevelt Library for aid when using the Morris L. Cooke papers. I would like to thank Mrs. Pearl W. Von Allmen of the School of Law Library of the University of Louisville for making it possible for me to use the Louis D. Brandeis papers. I owe thanks to David C. Mearns of the Manuscripts Division of the Library of Congress for permission to use the Frederick H. Newell and Otto S. Beyer papers, to Ralph H. Phelps for permission to use the American Engineering Council papers at the Engineering Societies Library in New York, and to John H. Moriarity for allowing me to use the Frank B. Gilbreth papers at Purdue. Thanks are also due to George P. Hammond of the Bancroft Library for access to the William Smyth collection and to Bruce Harding for assistance and hospitality while using the Ralph Mershon collection at Ohio State. I would like to give special thanks to Frederick Taft of the Sears Library of Case Western Reserve University for help in using the William E. Wickenden papers and for assistance on occasions too numerous to mention.

The publishers of three of my previous articles have generously allowed me to quote from the articles. Thanks are due to the Midcontinent American Studies Association, which published "Frederick Haynes Newell and the Revolt of the Engineers" in their *Journal*, III (Fall,

1962), 17–26; to the Trustees of the University of Pennsylvania, for permission to quote from "Veblen and the Engineers," *American Quarterly*, XIV (Spring, 1962), 64–72; and to John Wiley and Sons, Inc., for permission to quote "Science, Business and the American Engineer," which appeared in Robert Perrucci and Joel E. Gerstl, eds., *The Engineer and the Social System* (New York, 1969), 51–72.

The research for this book was assisted by two summer grants from Purdue University, while I was on the staff of that institution. Last, but certainly not least, I would like to thank those who have been closest to me in the preparation of this book. Mrs. Phyllis Garland typed the complete manuscript. Mrs. Marjorie Linn checked my quotations. And finally, I wish to thank my wife, Barbara, who not only read and criticized the manuscript, but also stood by me through the entire project from beginning to end.

LIST OF
ABBREVIATIONS

AAE	American Association of Engineers
AEC	American Engineering Council
AIEE	American Institute of Electrical Engineers
AIME	American Institute of Mining and Metallurgical Engineers
ASCE	American Society of Civil Engineers
ASME	American Society of Mechanical Engineers
ECPD	Engineers Council for Professional Development
EJC	Engineers Joint Council
FAES	Federated American Engineering Societies
IEEE	Institute of Electrical and Electronic Engineers
IRE	Institute of Radio Engineers
MMSA	Mining and Metallurgical Society of America
NELA	National Electric Light Association
NSPE	National Society of Professional Engineers
PROC	Proceedings
SAE	Society of Automotive Engineers
TRANS	Transactions

THE REVOLT OF THE ENGINEERS

1 · THE ENGINEER AND BUSINESS

The engineer is both a scientist and a businessman. Engineering is a scientific profession, yet the test of the engineer's work lies not in the laboratory, but in the marketplace. The claims of science and business have pulled the engineer, at times, in opposing directions. Indeed, one outside observer, Thorstein Veblen, assumed that an irrepressible conflict between science and business would thrust the engineer into the role of social revolutionary.[1]

While nothing like a soviet of engineers has appeared, the tensions between science and business have been among the most important forces shaping the engineer's role on the job, in his professional relations, and in the community at large. Veblen assumed that science and business made mutually exclusive demands on the engineer; in fact, however, they often complement one another. Both, for example, may benefit from technological progress. Nor is the existence of tensions necessarily detrimental to engineering work: attempting to resolve them may account for some of the engineer's drive and creativity.

Despite his mordant irony, Veblen was, in one sense, an optimist; he assumed that the tensions between business and science were resolvable, if only through a cataclysmic destruction of the former by the latter. In this he missed the essence of the engineer's dilemma which is, at base, bureaucracy, not capitalism. The engineer's problem has centered on a conflict between professional independence and bureaucratic loyalty, rather than between workmanlike and predatory instincts. Engineers are unlikely to become revolutionaries because such a role would violate the elitist premises of professionalism and because revolution would not eliminate the underlying source of difficulty. The engineer would still be a bureaucrat. Tensions with business have been dominant because in

1

the American context economic development has been carried out principally through the agencies of private capitalism. But engineers in government have experienced quite analogous conflicts, if anything more severe than those of privately employed engineers; and it can be argued that the market system, in providing a final test acceptable to both the engineer and his employer, has served to buffer discord between the two. Perhaps the engineer's problem ultimately is marginality: he is expected to be both a scientist and a businessman, but he is neither. A social revolution would merely alter the terms of his marginality without ending it.

The engineer's relation to bureaucracy is not new; he is the original organization man. The scientifically trained, professional engineer has characteristically appeared on the technical scene at the point of transition from small to large organizations. Economically, it makes little sense for small enterprises to employ engineers; the gains are not worth the costs. Large corporations, on the other hand, can more readily support engineers and research establishments, since they represent a small percentage of their total costs. Large corporations can get substantial net returns from rather small percentage gains in efficiency. Where large investments are at stake, the engineer can serve a useful function in eliminating guesswork and minimizing risks. Technically, large works are more likely to involve complexities than are small ones; and the larger the project, the more likely it is that such difficulties will transcend the capabilities of artisans and businessmen.

In the eighteenth and through much of the nineteenth century, America developed quite diverse and advanced technologies without requiring a corps of scientifically trained experts. The Mississippi River steamboat, for example, was developed in an era of relatively small, highly competitive enterprise by the cut-and-try methods of the practical mechanic, rather than by the rational analysis of scientists. It was large-scale organizations, such as the navy and private corporations, that supported experts who could approach the steam engine through the laws of thermodynamics. Similarly, it was not the small ironmasters who first called on the aid of science, but the corporate giants of the post-Civil War era. From the start, engineers have been associated with large-scale enterprises.[2]

There were two stages in the emergence of the engineering profession in America in the nineteenth century. The first demand for engineers came from the construction of large public works, such as canals and railroads, particularly in the period 1816 to 1850. The organizations that undertook these works were pioneers in technology and were also among the largest enterprises in America, representing aggregations of capital that were huge for their day. The civil engineering profession was called into being to meet the technical needs of these organizations.

In 1816, the engineering profession scarcely existed in America. It has been estimated that there were only about thirty engineers or quasi-engineers then available; but by 1850, when the census first took note of this new profession, there were 2,000 civil engineers. Canal and railroad construction generated not only the demand for engineers but, in large measure, the supply as well. From an early stage, organizations employing engineers found it convenient to group their technical staffs into a hierarchy of chief engineer, resident engineers, assistant engineers, and the like. Within this bureaucratic context, regular patterns of recruitment and training emerged on the job, and early engineering projects like the Erie Canal and the Baltimore and Ohio Railroad became famous as training schools for engineers. [3]

The rising demand for engineers by industry began the second stage in the emergence of the engineering profession. The golden age for the application of science to American industry came from 1880 to 1920, a period which also witnessed the rise of large industrial corporations. In these forty years, the engineering profession increased by almost 2,000 percent, from 7,000 to 136,000 members. The civil engineer was over-shadowed by the new technical specialists who emerged to meet the needs of industry: by the mining, metallurgical, mechanical, electrical, and chemical engineers. The astonishing growth of engineering continued, though at a less rapid rate, after 1920. In 1930, there were 226,000 engineers when the depression put a brake on expansion; by 1940, the number was but little higher—260,000. Postwar prosperity increased the size of the profession past the half-million mark in 1950, and to over 800,000 by 1960. Engineering is by far the largest of the new professions called forth by the industrial revolution. [4]

The rise of the engineering profession was accompanied by a scientific revolution in technology. The change was not a sudden one; in most cases engineering built upon and extended traditional techniques. But professionalism was associated with a slow incorporation of scientific methods and theory into technology and the accumulation of an esoteric body of technical knowledge. Professionalism was a means of preserving, transmitting, and increasing this knowledge. The transition from traditional rule-of-thumb methods to scientific rationality constitutes a change as momentous in its long-term implications as the industrial revolution itself.

The development of engineering education constitutes a sensitive indicator of the shift from art to science in technology. The early civil engineers were educated by self-study and on-the-job training. Only a minority received a college degree. By 1870, there were twenty-one engineering colleges, but only 866 degrees had been conferred. College education became increasingly common after 1870. By 1896, there were 110 engineering colleges. The number of students increased rapidly, from 1,000 in 1890 to 10,000 in 1900. Only with the twentieth century, however, did the college diploma become the normal means of admission to engineering practice. [5]

College training was a sign of a greater emphasis on science. But even here the change was evolutionary. The engineering curriculum of the latter nineteenth century placed as much or more emphasis on craft skills as upon scientific training. In 1875, Alexander L. Holley, a prominent mechanical engineer, argued that all the great engineering triumphs of his day owed more to art than to science. The aim of college education, Holley maintained, was to train "not *men* of good general education, but *artisans* of good general education," for as Holley emphasized, "the art must precede the science."[6] Despite an increasing emphasis on science, engineering educators down to 1920 seriously debated whether engineering students ought to learn the calculus. Some of them thought such courses were "cultural" embellishments to the curriculum. Only since the end of the Second World War has the balance in engineering education shifted unequivocally toward science. [7]

The origin of engineers carries with it built-in tensions between the bureaucratic loyalty demanded by employers and the independence implicit in professionalism. It is important to note, however, that these tensions are not, as Veblen thought, the outgrowth of a clash between the rationality of science and the irrationality of capitalism. The scientific knowledge possessed by the engineer is highly rational, but his professionalism derives from the mere possession of esoteric knowledge, not its specific content. Incomprehensibility to laymen, rather than rationality, is the foundation of professionalism. In essence, the professional values adopted by American engineers are the same as those of other professions. They may be summarized under the headings of autonomy, colleagual control of professional work, and social responsibility. [8]

Perhaps the most invariant demand by all professions is for autonomy. The classic argument is that outsiders are unable to judge or control professional work, since it involves esoteric knowledge they do not understand. Autonomy operates on at least two distinct levels: it applies to engineers in their corporate sense as an organized profession and to the individual engineer in relation to his employer. In both cases, conflicts have appeared between business demands and the ideal of professional independence. Businessmen usually concede that engineering societies should be free of external control, but in practice business domination is not uncommon. Employers have been unwilling to grant autonomy to their employees, even in principle. They have assumed that the engineer, like any other employee, should take orders. Some engineers, however, have maintained that the engineer, like the doctor, should prescribe the course to be followed and that the very essence of professionalism lies in not taking orders from an employer.[9] The employer, of course, has the power to reward and punish. But the value of the engineer generally hinges on his being a professional in the sense of being both a master of an ever-growing body of knowledge and a creative contributor to that knowledge. Such men are the ones most likely to be inspired by professional ideals. As a result, the role of the engineer represents a patchwork of compromises between professional ideals and business demands.

The argument for colleagual control of professional work is closely related to that for autonomy. Since professional work cannot be understood fully by outsiders, the person in charge of such work should be a member of the profession. In this manner, doctors have insisted that the heads of hospitals and medical schools should be members of the medical profession. In the same vein, engineers have mantained that engineers should be in charge of engineering work.[10] In practice, engineering departments are usually headed by engineers. But in the case of engineering, this principle can be extended much further. Engineering is intimately related to fundamental choices of policy made by the organizations employing engineers. This can lead to the assertion that engineers ought to be placed in command of the large organizations, public and private, which direct engineering. This is tantamount to saying that society should be ruled by engineers. A more representative manifestation of the ideal of colleagual control has been the repeated demand by the profession for reform of governmental public-works policy. Engineers have advocated the creation of a cabinet-level department of engineering to be headed by a civilian engineer.[11]

Although the arguments for autonomy and colleagual control are fundamentally similar, one of the basic dilemmas of modern engineers has been that these two goals are not completely compatible. Organizations like the federal government or a modern corporation have other ends in view than the best and most efficient engineering. Doctors are in a more fortunate position, since it may be assumed that the professional aim—health—is identical to the ends of the large organizations employing doctors, such as hospitals and medical schools. But unlike medicine, engineering serves purposes ulterior to itself. Colleagual control of engineering implies, in the extreme, a change in basic social values, to make those of engineering supreme. Such aspirations open up the possibility that outside organizations might reciprocally seek to control engineering, as something potentially dangerous to their purposes.

Professional freedom implies social responsibility. The professional man has a special responsibility to see that his knowledge is used for the benefit of the community. Social responsibility points in two directions: inwardly, at self-policing to prevent abuses by colleagues, and outwardly, to the making of public policy. In either case, it is with social responsibility that professionalism comes most clearly into conflict with bureaucracy. This may be seen in two possible meanings of the term "responsibility." On the one hand, there is the bureaucratic sense implied in the phrase "responsible public official." This denotes responsibility in executing policies, but not necessarily in formulating them. On the other hand, the term "professional responsibility" entails an independent determination of policy by the professional man, based on esoteric knowledge and guided by a sense of public duty. An assertion of professional responsiblity, therefore, may signify a rejection of bureaucratic authority. In this manner, the scientists' crusade against the May-Johnson Bill for postwar control of atomic energy was both an assertion of a professional responsibility and a rebellion against General Groves and the formal hierarchy of the Manhattan Project.

For engineers, the most overt element of professionalism has been an obsessive concern for social status. Although the income and the power of engineers would of course be enhanced by professionalism, these ends have been given second place, at least verbally. Professionalism has induced engineers to seek greater deference, in particular, to gain the same social recognition accorded to the traditional learned professions, law and medicine. Spokesmen for the engineering profession have, in fact, frequently made status the fundamental aim, and other professional values means to this end. Thus, engineers have argued that in order to gain more status their profession should show a greater sense of social responsibility. [12]

Although engineers emphasize the importance of status, it is not clear that this distinguishes their goals from those of other interest groups. Engineers differ from nonprofessional groups chiefly in that they are more likely to rationalize their ambitions in terms of protecting the public. Following the cue of the older professions, engineers have secured the enactment of licensing laws, and they have endeavored to raise standards in education and practice. Such measures enable professions to limit competition and alter supply-and-demand relationships in their favor. Engineers have hoped to achieve in this way many of the same goals sought by labor unions through such devices as the closed shop and strike. Professional autonomy and control of the profession's work by colleagues, if fully realized, would lead to control of the conditions of work, an end pursued equally, if by other means, by labor unions. As an interest-group strategy, professionalism offers several advantages. It is "dignified," since the professional abjures "selfish" behavior, at least verbally, and gains group advantage on the pretext of protecting the public. Professionals emphasize the intimacy of the personal relationship with clients, rather than the cash nexus between buyer and seller, employer and employee.

Professionalism for engineers is not exclusively a matter of esoteric knowledge. Engineers do not seek autonomy simply because they are professionals; to some extent they have adopted professionalism as a way of gaining autonomy. Professionalism was, in part, a reaction against organization and bureaucracy. It was a way to prevent engineers from becoming mere cogs in a vast industrial machine. Thus, in 1939, Vannevar Bush made an eloquent plea for a professional spirit in engineering. Without this spirit, Bush thought,

> we may as well resign ourselves to a general absorption as controlled employees, and to the disappearance of our independence. We may as well conclude that we are merely one more group of the population . . . forced in this direction and that by the conflict between the great forces of a civilized community, with no higher ideals than to serve as directed. [13]

Professionalism carries overtones of elitism that grate against the egalitarian assumptions of American democracy. Professionalism stresses hierarchy and the importance of the expert; its emphasis is on the creative few, rather than on the many. But professionalism, not itself democratic, may serve democratic ends. It is one means of preserving the ideal of the autonomous individual, without which democracy could scarcely exist. Democracy requires not only freedom, but an informed public opinion. One of the problems of the modern age is that many issues of public policy involve technical matters. Independent and informed judgments of these questions are badly needed by the public. Professionalism, because of its stress on social responsibility, offers one way of meeting this need by establishing a legitimate role for private judgment by engineers, protected and encouraged by an organized profession.

Whether the emphasis be on esoteric knowledge, public service, or selfish interest, professionalism requires that engineers identify themselves with their profession. Engineers must think of themselves as engineers before they can constitute a profession. There are several factors that link engineers together as a group and encourage self-consciousness. As with all professions, the fundamental tie is a common body of knowledge. Self-interest is another powerful cohesive force. Both find a natural focus in the professional society, which brings engineers together and gives them a sense of corporate identity. These factors have helped to produce a steady push toward professionalism among engineers.

Professionalism, however, has met powerful resistance from business. Business has been reluctant to grant independence to employees. The claims of professionals rest fundamentally on esoteric knowledge; to reduce the importance of this knowledge is to weaken the engineers' aspirations for autonomy. To some extent this may take the form of a depreciation of "theory" and an emphasis on practicality. But if businessmen tend to depreciate esoteric knowledge, they cannot wish it away. Experts have been of increasing, not diminishing importance. A more pervasive and effective argument is the priority of business needs over technical considerations.[14] This contention is not without force. The transition from art to science in engineering has been slow and partial. Even where technical knowledge has achieved a high state of perfection, its importance is limited by the exigencies of business. Engineers work in complex organizations. Engineering is only one factor among many that contribute to their success.

In the long run, the most effective check on professionalism by business has been a career line that carries most engineers eventually into management. The promise of a lucrative career in business does much to ensure the loyalty of the engineering staff. Conversely, it undermines engineers' identification with their profession. Social mobility carries with it an alternative set of values associated with the businessman's ideology of individualism. These values compete with, and to some degree conflict with, those of professionalism. Thus, professionals stress the importance of expert knowledge, but businessmen stress the role of personal characteristics, such as loyalty, drive, initiative, and hard work. Professions value lifetime dedication. But business makes engineering a phase in a successful career rather than a career in itself. Insofar as business treats engineering merely as a stepping stone to management, it represents a denial of much that professions stand for.

It is possible to distinguish several stages in the typical engineering career. The most important source for understanding the engineer's background is a study conducted by the Society for the Promotion of Engineering Education in 1924, as part of a large survey of engineering education. It was based on a questionnaire administered to 20 percent of all engineering freshmen admitted in the fall of that year, 4,079 students selected from thirty-two engineering schools so as to constitute a representative cross section. A very large proportion of the parents of these students were members of the old middle class: 42.5 percent were owners or proprietors of businesses. Of the remainder, more than one-quarter were members of the new middle class: 28.2 percent were employed in executive or supervisory positions and 5.6 percent were engineers or teachers. Another 13 percent of the total were skilled workers, but only 2.7 percent were unskilled workmen and 3.5 percent were clerks.[15]

The engineering students surveyed by the 1924 study were drawn from the poorer and less well-educated segments of the middle class. Almost all of the parents who were owners or proprietors were engaged in small mercantile enterprises or farming. Only 13 percent of the fathers had a college degree. Forty percent had graduated from high school, and the same number had a grammar school education or less. Sixteen percent had started but not finished high school. That the parents were not especially well-off is further suggested by the fact that 90 percent of the freshmen had to work a year before starting college.[16]

The parents of these students were overwhelmingly of Anglo-Saxon or northwestern European stock. Of the students, 96.2 percent were native-born, as were 73.6 percent of their parents and 60.7 percent of their grandparents. Of the grandparents not born in the United States or Canada, two out of three were from northwestern European countries. Only 10 percent of all the grandparents were from Latin, Slavic, or other countries.

More than three out of five of the students' families lived in small towns, villages, or on farms. Only 38.7 percent came from cities with a population of over 25,000.[17] A recent study of similar scope indicates that the proportion of engineers drawn from families of blue-collar workers has significantly increased. Presumably this has affected the ethnic background of the engineers concerned. But the large city continues to be underrepresented in the profession.[18]

The selectivity in recruitment of engineering students from middle-class, small-town, and old-stock families goes far to account for the profession's strong commitment to traditional individualism. One engineer, in tracing his own faith in rugged individualism to the self-reliance of his father, suggested that his own experience was typical of his generation.[19] Another engineer, commenting on the frontier individualism of his home community, noted that "no one growing up in such an environment could escape being influenced by it."[20] A frequent theme in discussions of success in engineering is the importance of an "inner drive," or an "inner urge," or more simply of initiative.[21] It is perhaps no accident that the term "rugged individualism" was given currency by an engineer, Herbert Hoover. But predisposition to individualism also provides the foundations for the development of a commitment to business.

Engineering education is susceptible to business influence in a number of ways. Businessmen serve as trustees of colleges, on alumni boards, and on committees of technical societies concerned with technical education.[22] Engineering educators suffer from the same divided loyalties as other engineers, and some of them have been important spokesmen for a business point of view. Many businessmen, and some engineering educators, have assumed that the buyer, business, has the right to determine both the technical skills to be taught and the ideas to be implanted in students' minds. Business demands on this score have ranged from training in "sound economics" to concern for international responsibilities.[23] In 1932, a group of engineering educators at Ohio State University became interested in training engineering students in social responsibilities; but before undertaking anything definite, they prudently sampled the opinions of business leaders on major issues of the day. One of the educators noted that the "industrial system" was not then working well. He reflected:

> What would happen if these socially awakened young men should come to the conclusion that its very fundamentals are wrong. . . . Manifestly such young reformers might find a cold reception in industry. . . . It is therefore well to ascertain what leading executives think about such matters.[24]

Despite strong business influence, however, the feeling of professional independence is probably stronger in engineering schools than elsewhere. Educators tend to think of themselves as independent practitioners. Shielded by traditions of academic freedom and in contact with other professions, engineering educators have been in a position to assert a professional commitment. It is not surprising that educators have been prominent in movements to uphold and advance professionalism. Perhaps the educators' most important impact has been on their students whom they inculcate with professional ideals.

The next stage in the engineer's development is that of professional engineering proper. The transition is quite marked geographically and environmentally, since most engineers leave their home states on completion of college—59 percent, as against an average of only 38 percent for all other college graduates.[25] The young graduate does not qualify immediately as a full-fledged professional. Four years at college are barely sufficient to lay the foundations upon which the young man must continue to build for a lengthy period thereafter. This process is not simply one of absorbing more knowledge; managerial and personal skills come into play also. The final judge of the young man's gradual advance is as much or more his employer as his professional colleagues. Major engineering societies recognize four stages in the engineer's development, which they embody in their grades of membership: from student to junior, to associate, and finally to full member. But most such societies place as one of their requirements for the higher grades of membership that the candidate be in "responsible charge" of engineering work, a test as much of his bureaucratic status and business success as of his engineering knowledge. Success in business overlaps and may eclipse success in the engineer's profession.

Most engineers work in industrial bureaucracies, which are capable of exerting a considerable amount of pressure on the individual. The effectiveness of this influence is heightened when the individual seeks not only to keep his job, but to rise in the hierarchy. The price of success, in at least some cases, is total and undivided loyalty. An engineering educator made this point in 1935 when he asked a group of young engineers:

> Are you truly loyal to your employer's interests?
> Do you try to advance these on every opportunity?
> Do you have a fighting spirit for the reputation of 'our company' and 'our products'? The merging of your employer's interests into your own is one of the surest signs of real progress in business life.[26]

Conformist pressures on the job are not limited to company loyalty; they extend to all sorts of social and political norms. As one engineering society president, J. F. Coleman, noted, the employed engineer cannot be as open in expressing his opinion on socio-economic subjects as one who is a free agent.[27] One young engineer complained, in 1942, that on the job "everything possible was done to make alert youth conform to routine. Ideas were discouraged, social and political interests frowned upon."[28] Pressure appears to be especially great against heterodox ideas. A young mechanical engineer wrote, in 1941, that many engineers favored joining unions, but that they were "under the terror of jeopardizing their connections by coming out into the open with their opinions."[29] Another engineer wrote to a leading engineering journal, in 1940, "to my mind for individual engineers themselves to have published in *Mechanical Engineering* their own views of what they would like is very likely to bring about some action which may put us all on the spot."[30] A distinguished civil engineer, Daniel W. Mead, in 1929 advised young engineers to avoid unorthodox dress, since "conspicuous dress usually shows unfortunate idiosyncrasies which need to be eliminated."[31]

Employers have occasionally urged engineers to show more social responsibility. But by this they generally have in mind the defense of business, rather than independent action by the profession. Engineering support for business is often taken for granted. As one mining engineer told a group of younger engineers in 1931:

> We as employees owe the company our best efforts not only while at work but at all times, promoting always the company's interests by representing them favorably. This we can best do by active, harmonious participation in the community social, civil, and spiritual activity.[32]

He thought the engineer could best serve his employer by being a "booster imbued with the spirit of boosting at all times."[33] H. B. Gear, an engineer who became the vice-president of the Commonwealth Edison Company of Chicago, in 1942 maintained that engineering best served society as an "implement of management."[34] It is not clear whether Gear had in mind the sort of social service provided by Samuel Insull, the founder of Commonwealth Edison, but in any case Gear's idea of the preeminence of business considerations left little room for an assertion of social responsibility by an independent profession.

The coercive element within bureaucracies should not be exaggerated, however. Pressures internalized within the individual are far more effective. One of the most acute observers of the profession, William E. Wickenden, noted that engineers seeking social premises look to "the presuppositions of the man higher up who sets the engineer at his problem and reserves final judgment on his recommendations."[35] An ardent advocate of professional values, Arthur E. Morgan, sadly noted that "the engineer tends to reflect, somewhat uncritically, the social attitude of his employer."[36] He implored engineers to abandon their "moral servitude" and develop an independent philosophy.[37]

The final phase of the engineer's development, not achieved by all, is the transition from professional engineering to business management. A survey of the six largest American engineering societies, in 1946, revealed that over one-third of their membership was engaged in management.[38] A study of 5,000 engineering graduates by the Society for the Promotion of Engineering Education, in the 1920s, suggested that, on the basis of their sample, over 60 percent of all engineers made this shift eventually. There was a secondary drift from technical to nontechnical management and administration. Half had made this change by forty years of age, and the proportion increased to three-quarters after fifty-five.[39] No doubt monetary rewards have played an important role in encouraging engineers to move into management. Several studies of engineers have stressed the large premiums paid for managerial work as against purely technical engineering.[40] One result of this trend has been that engineers constitute a large and increasing segment of the upper echelons of business. Mabel Newcomer found that the proportion of engineers in the managerial elite of presidents and board chairmen grew from one in eight in 1900 to almost one in five in 1950.[41]

It is not uncommon for engineers who have risen into managerial positions to think of themselves as businessmen. John Mills, a discerning observer of his profession, noted that as engineers climb into supervisory positions, they tend to be blinded by visions of future promotions and to lose sight of colleagues lower down.[42] Although many such men no longer think of themselves as engineers, they may retain professional-society membership for business reasons. But professional commitment is clearly diminished. Others sever their ties with engineering completely. The 1960 census, which relied on the self-identifications of respondents for occupations, counted 541,000 engineers and scientists in manufacturing, the bulk of whom were engineers. A follow-up survey by the National Science Foundation discovered an additional 73,000 technically qualified persons who had been missed because they had identified themselves with nontechnical occupations, presumably managerial.[43]

A number of engineers have looked to the major professional associations to offset the power of business. Arthur E. Morgan, the first head of the Tennessee Valley Authority, was deeply concerned by the fact that the engineer employed in a bureaucracy "tends to be not a free agent, but a technical implement of other men's purposes." But he pessimistically concluded that "the lone individual is largely helpless."[44] Morgan, however, hoped that engineering societies might enable their members collectively to gain the freedom they lacked as individuals. He was aware this would not be easy; he pointed out:

> For engineering associations to be a democratic force in the public interest, they must be more than the summing up of the selfish interests of their members, and more than the reflections of the views of their several employers. . . . Otherwise an engineering organization may be only a more powerful means of regimenting the members along lines of policy which their employers happen to be following.[45]

The professional ideal expressed by Morgan was difficult to attain because business influence had penetrated the very citadel of professionalism, the engineering society. The most important agents of business influence within the organized profession are engineers who have moved into management; they constitute a large and powerful minority of senior and successful men. As the editor of *Mining and Metallurgy* observed: "naturally such men are highly influential in establishing the policies of engineering societies."[46]

Engineers who have risen into top management present a dilemma for engineering societies. Although it has been generally conceded that engineers engaged in technical management are still engineers, this is far from clear in the case of those who have gone up the corporate hierarchy to positions involving general management and administration. Some engineers have argued that such men are no more practicing their profession than a lawyer in a similar position is practicing his.[47] This argument implies that such men should be excluded from the higher grades of membership of engineering societies and, hence, from effective control of their profession. Other engineers have taken a contrary position. They have maintained that the engineer in top management has fulfilled the engineer's cherished ideal of success and that even those who have lost any active interest in technical matters should be encouraged to join engineering societies and participate in their government. In 1953, Frederick S. Blackall, Jr., the president of the American Society of Mechanical Engineers, maintained that

> if ASME . . . is one of the most important factors in moulding the character and competence of the technical staff, then too, most certainly the men of top management should make it their

business to have a voice in its affairs. Here will be found a splendid sounding board for their views and platforms for their leadership.[48]

However, the fact that engineers in management tend to lose their identification as engineers presents a further difficulty. In 1927, for example, Blackall's society elected Charles M. Schwab as its president; yet in his *Who's Who* autobiography, Schwab listed himself as a "capitalist," down to 1934, and as a "steel manufacturer" thereafter; he did not mention his presidency of the American Society of Mechanical Engineers or even his membership in that organization.[49]

To some extent business influence is checked by the power of majority rule. But engineering societies are not perfectly functioning democracies, and there are serious limitations to the control that rank-and-file members can exercise. Some societies long denied younger members the right to vote; even without formal restrictions on voting, the idea of a hierarchy of professional excellence implicit in membership grades tends to give power to the senior full members and to deny significant influence to junior members. In certain specific instances, employers have canvassed their technical staffs to line-up votes on particular issues. But more commonly, widespread ignorance and apathy give inordinate power to comparatively small minorities. Most members live in remote parts of the country and are little interested in society affairs. Nominations are controlled by committees whose actions are usually shrouded in secrecy; the election presents a mere formality.[50]

It is considered bad form to publicize the inner workings of engineering societies. Even in those rare cases where elections have been contested, both parties have usually taken precautions to keep the real issues secret. In 1914, Morris L. Cooke led a revolt against excessive dominance of the American Society of Mechanical Engineers by the utilities. His election to the society's governing board was opposed by a rival candidate put up by the utility interests. Not only did no mention of this appear in the ASME's official publications, but it was deliberately obscured in the informal circulars distributed by both sides. The opposition based its position on a technicality in the manner of Cooke's nomination. Cooke, though a foremost advocate of publicity in engineering-society affairs, also avoided the real issue for tactical reasons. In writing to a close friend concerning his circulars, Cooke noted that "the utility matter is kept in the background and can be used with telling effect, but we do not draw on ourselves the charge of washing the society's soiled linen in the public."[51]

Given the secrecy surrounding engineering-society affairs and the indifference of most members, the machinery of society government offers many opportunities for small but active groups to exercise control. Three committees have been particularly important as sources of business influence. They are those concerned with nominations, new members, and publications. By controlling nominations, a group can make sure that engineers sympathetic with its interests are in strategic positions in the society's government. Control of membership, in the long run, is the most important power of all, since this committee does much to determine the future composition of the society. Control of membership committees can also be employed to keep out of a particular society individuals not acceptable to particular business interests. Edward W. Bemis, who often took the public side in utilities-valuation cases, was advised not to apply for membership in the American Society of Mechanical Engineers because of the power of the utilities in the committees concerned with passing on new membership applications.[52]

Perhaps the most influential committees of engineering societies are those concerned with publications. Such committees can control the publication of technical papers, censor heretical opinions, or silence proposals that might embarrass particular business interests. The electric utilities have been especially active and successful in this respect. Their power in the American Institute of Electrical Engineers led to a prohibition of papers dealing with costs and rate making. According to Governor Pinchot of Pennsylvania, this practice seriously hampered the efforts of public agencies to regulate utilities.[53] Other societies, lacking such a prohibition, have been prevented from publishing papers opposed by utilities interests. In May of 1938, for example, Gregory M. Dexter submitted a paper to the American Society of Mechanical Engineers that dealt with electric rates of a subsidiary of Consolidated Edison in Scarsdale, New York. A long series of delays and rewritings ensued. Dexter claimed he would be asked to write it one way, and then he would find the "rules" had been changed and he would have to write it another way. Not until January, 1942, was the paper finally rejected. On the committee that ruled the paper unacceptable were several engineers connected in various ways with utilities interests, among whom was an advisory member who held the position of assistant engineer for Consolidated Edison.[54]

Virtually all of the major engineering societies in America have had more or less celebrated cases of censorship, involving a broad spectrum of business interests. Within the American Institute of Mining Engineers, for example, the representatives of the coal and oil industries have been accused of censorship. In 1916, W. H. Shockley claimed that a paper of his was suppressed at the instance of the anthracite section because it presented government statistics showing the wages of coal miners were inadequate.[55] In 1922, another member, Robert B. Brinsmade, complained that a paper of his was censored because of certain comments about oil policy. He sarcastically suggested that his letters would enable members to avoid opinions "which have been put on the prohibited index by the august Committee of Publications."[56]

This sort of censorship is made more effective by the codes of ethics adopted by some societies. These codes may prohibit criticism of fellow engineers and the discussion of engineering subjects in the general press. The codes of ethics of the American Institute of Electrical Engineers and the American Society of Mechanical Engineers both provided that "technical discussions and criticisms of engineering subjects should not be conducted in the public press, but before engineering societies, or in the technical press."[57] These codes of ethics are seldom enforced, so the sanction is perhaps limited. But it is not trivial, either. In order to circumvent the restriction on publication outside the society, Frederick W. Taylor had one of his works privately printed and gave a free copy to each member of the American Society of Mechanical Engineers prior to general circulation.[58] This expedient, however, was available only to those with considerable means.

Censorship, though not absolute, is a significant deterrent to free publication by engineers. In 1932, it led to the expulsion of Bernhard F. Jakobsen and James H. Payne from the American Society of Civil Engineers. Payne and Jakobsen, in 1930, had exposed certain malpractices in connection with the construction of a dam for Los Angeles County. They published their findings in the pages of a local newspaper, and they were critical of the chief engineer, among others. The two engineers who made the exposure were then expelled from the American Society of Civil Engineers for unprofessional conduct. It was later established that the contractor had bribed the chairman of the Board of Supervisors, who was sent to jail. The contractor was forced to return over $700,000 to the county. Despite a plea from the Los Angeles County Board of Supervisors to reconsider, the ASCE refused to reinstate Jakobsen and Payne.[59]

In many matters, such as censorship, engineers who hold managerial positions have been the principal agents of business influence in engineering societies. But they are by no means the only source of business power within the profession. Even if such men were excluded from the higher grades of membership—a favorite remedy for some reformers—

business influence would remain substantial. Virtually all American engineering societies are financially dependent on business. This is clearest in those smaller societies that have company members; here the subsidy is direct.[60] Other engineering societies are nominally supported solely by the dues of members. But in fact, they receive indirect financial support from business. A substantial number of members have their dues paid by their employers. A survey in 1947 revealed that, in the sample studied, some 30 percent of the employers regularly paid the dues of some of their employees in certain societies.[61] In 1940, a spokesman for the membership committee of the American Institute of Mining Engineers complained that qualified men were not joining the institute because they were waiting for their employers to pay their dues; he found it necessary to remind them that membership in the institute was a personal matter.[62]

A second form of indirect business subsidy is that of employers paying traveling expenses and allowing time off for employees to attend engineering-society meetings, especially when presenting a paper, serving on a committee, or participating in a discussion. The official journal of the American Society of Mechanical Engineers, *Mechanical Engineering*, commented editorially that engineering societies must accept such aid, since without it they "would have to shut up shop, as a majority of the work of engineering societies is done by its members who are encouraged by their employers to do this work."[63]

Business subsidies pose a delicate problem for engineers. Some societies have openly admitted the practice and defended it. *Mechanical Engineering* urged employers to look on engineering-society activities of an employee as an "assignment" and suggested that the employer "send him to a meeting, or encourage his participation in some other form of society activity, with the feeling that he is representing his company's as well as his own interests."[64] Other engineers, concerned with professional independence, have argued that such practices are harmful. One engineer maintained that subsidies from business made their recipients unfit to represent the engineering profession, and he urged support of the National Society of Professional Engineers since it was "for the benefit of the engineering profession primarily and not for the advancement of corporation interests."[65] A further problem here is that "pure" professional organizations, such as the NSPE, do not publish technical papers. It would be financially impossible for them to do so, even if employers would permit publication of papers by their employees by an organization outside their influence.

Although there are significant areas of conflict between business and professionalism, the dimensions of the clash should not be exaggerated. Neither could exist under present circumstances without the other. Modern business needs highly esoteric technical knowledge, and only professionals can supply it. Technologists need organizations in order to apply their knowledge; unlike science, technology cannot exist for its own sake. The problem has been to find suitable mechanisms of balance and accommodation. One of the basic problems of American engineers is that the balance has tended to shift too far in the direction of business, and accommodation has taken place largely on terms laid down by employers. The professional independence of engineers has been drastically curtailed. The losers are not just engineers. The public would benefit greatly from the unbiased evaluations of technical matters that an independent profession could provide. American business too might profit in the long run from the presence of a loyal opposition.

NOTES

1. Thorstein Veblen, *The Engineers and the Price System* (New York, 1933), 70–76.

2. Louis C. Hunter, *Steamboats on Western Waters* (Cambridge, 1949), 175–180, 307–309; William F. Durand, *Robert Henry Thurston* (New York, 1929), 201, *passim;* W. Paul Strassman, *Risk and Technological Innovation: American Manufacturing Methods During the 19th Century* (Ithaca, 1956), 28–46.

3. Daniel Hovey Calhoun, *The American Civil Engineer, Origins and Conflict* (Cambridge, 1960), 22, 27–29, 48–50.

4. U.S. Bureau of Census, *Sixteenth Census, Population, Comparative Occupational Statistics for the United States, 1870–1940* (Washington, 1943), p. 111, table 8; Jay M. Gould, *Technical Elite* (New York, 1966), 172. Here and elsewhere I have rounded off figures for total number of engineers to the nearest thousand.

5. Society for the Promotion of Engineering Education, *Report of the Investigation of Engineering Education, 1923–1929,* 2 vols. (Pittsburgh, 1934), I, 541–547 (hereafter cited as SPEE, *Report of Investigation*), and David L. Fiske, "Are the Professions Overcrowded?" *Civil Engineering,* IV (June, 1934), p. 16, table I.

6. Alexander L. Holley, "The Inadequate Union of Engineering Science and Art," *Trans AIME,* IV (1875–1876), 191–192, 201.

7. C. R. Mann, "Report of the Joint Committee on Engineering Education," *Engineering Education,* IX (September, 1918), 19–26; H. D. Gaylord, "The Relation of Mathematical Training to the Engineering Profession," *Engineering*

Education, VII (October, 1916), 54–55; Frank McKibben, "The Colleges and the War," *Engineering Education*, IX (May, 1919), 363; and George F. Swain, "The Liberal Element in Engineering Education," *Engineering Education*, IX (December, 1918), 97–107.

8. The literature on professions is vast, but I have been particularly influenced by the writings of Everett C. Hughes. See Everett C. Hughes, "Professions," in Kenneth S. Lynn, ed., *The Professions in America* (Boston, 1965), 1–14, and Everett C. Hughes, *Men and Their Work* (Glencoe, Illinois, 1958). For a discussion of professionalism by an engineer, see William E. Wickenden, "Toward the Making of a Profession," *Electrical Engineering*, LIII, pt. 2 (August, 1934), 1146–48.

9. Frederick H. Newell, "A Practical Plan of Engineering Cooperation," *Journal of the Cleveland Engineering Society*, IX (March, 1917), 311.

10. For an example, see Hunter McDonald, "Address at the Annual Convention," *Trans ASCE*, LXXVII (December, 1914), 1755.

11. For examples, see Clemens Herschel, "Address at the Annual Convention," *Trans ASCE*, LXXX (1916), 1307–14, and "News of the Federated American Engineering Societies," *Mechanical Engineering*, XLIII (June, 1921), 421–422.

12. C. O. Mailloux, "The Evolution of the Institute and of Its Members," *Trans AIEE*, XXXIII, pt. 1 (January–June, 1914), 827–834.

13. Vannevar Bush, "The Professional Spirit in Engineering," *Mechanical Engineering*, LXI (March, 1939), 198.

14. For one engineering educator's reaction to demands for "practicality," see Abraham Press, "Education Versus Engineering," *Engineering Education*, VIII (September, 1917), 28–30. For an example of the stress on business needs, see H. B. Gear, "The Engineer as an Implement of Management," *Electrical Engineering*, LXI (August, 1942), 426–427.

15. SPEE, *Report of Investigation*, I, 161–164, 188.

16. *Ibid.*, 164–166, 189.

17. *Ibid.*, 162–163, 188–189.

18. Robert Perrucci, William K. Le Bold, and Warren E. Howland, "The Engineer in Industry and Government," *Journal of Engineering Education*, LVI (March, 1966), 239–240.

19. A. W. Robertson, "Industry's New Responsibilities," *Electrical Engineering*, L (September, 1931), 719.

20. Frank B. Jewett, "An Exciting and Pleasant Forty Years," *Electrical Engineering*, LVIII (April, 1939), 161.

21. William E. Wickenden, "The Young Engineer Facing Tomorrow," *Mechanical Engineering*, LXI (May, 1939), 347, 348, and John C. Parker, "Responsibilities in the AIEE," *Electrical Engineering*, LVIII (August, 1939), 331–332.

22. "Industry's Influence," *Mechanical Engineering*, LXVII (August, 1945), 499–500.

23. Thorndike Saville, "Engineering Education in a Changing World," *Journal of Engineering Education*, XLI (September, 1950), 5. See also L. W. W. Morrow, "Industry Demands and Engineering Education," *Electrical Engineering*, LIII, pt. 1 (April, 1934), 518–522.

24. A. Norman, "Industrial Fundamentals," *Journal of Engineering Education*, XXII (March, 1932), 537.

25. Ernest Havemann and Patricia Salter West, *They Went to College* (New York, 1952), 235.

26. A. G. Christie, "Engineers' Business Contacts," *Mechanical Engineering*, LVII (February, 1935), 88.

27. J. F. Coleman, "Reflections on the Status of the Engineer," *Trans ASCE*, XCIV (1930), 1346.

28. Walter J. Gray, "Present-Day Responsibilities of the Engineer" (letter to the editors), *Civil Engineering*, XII (December, 1942), 687.

29. Andrew A. Bato, "Unionization of Engineers" (letter to the editors), *Mechanical Engineering*, LXIII (June, 1941), 476.

30. James M. Sherilla (letter to the editors), *Mechanical Engineering*, LXII (May, 1940), 412.

31. Daniel W. Mead, "The Engineer and His Education," in Dugald C. Jackson and W. Paul Jones, eds., *The Profession of Engineering* (New York, 1929), 28.

32. Henry Coleman, "Loyalty," *Mining and Metallurgy*, XII (August, 1931), 374–375.

33. *Ibid.*

34. H. B. Gear, "Engineering as an Implement of Management," *Electrical Engineering*, LXI (August, 1942), 426–427.

35. William E. Wickenden, "The Social Sciences and Engineering Education," *Mechanical Engineering*, LX (February, 1938), 149.

36. Arthur E. Morgan, "The Faith of the Engineer," *Civil Engineering*, XII (August, 1942), 421.

37. Arthur E. Morgan, "Engineer's Share in Democracy," *Civil Engineering*, IX (November, 1939), 638.

38. Andrew Fraser, *The Engineering Profession in Transition* (New York, 1947), p. 45, table 3.5g. The societies were the American Society of Civil Engineers, the American Institute of Mining Engineers, the American Society of Mechanical Engineers, the American Institute of Electrical Engineers, the American Institute of Chemical Engineers, and the National Society of Professional Engineers.

39. SPEE, *Report of Investigation*, I, 231–232. See also William E. Wickenden and Eliot Dunlap Smith, "Engineers, Managers, and Engineering Education," *Journal of Engineering Education*, XXII (June, 1932), 846–847.

40. Fraser, *Engineering Profession in Transition*, p. 28, table 1.9a. See also "1930 Earnings of Mechanical Engineers," *Mechanical Engineering*, LIII (September, 1931), 655.

41. Mabel Newcomer, *The Big Business Executive* (New York, 1955), 90. In absolute terms, this increase has been still larger since the group of top executives grew from 284 in 1900 to 319 in 1925, and to 868 in 1950. Jay M. Gould has recently extended Newcomer's figures to 1964. He found that the proportion of engineers and scientists had increased from about one in five in 1950 to about one in three in 1964 (see his *Technical Elite*, 82–84).

42. John Mills, *The Engineer in Society* (New York, 1946), 116, 138–139.

43. Gould, *Technical Elite*, 130.

44. Arthur E. Morgan, "Engineer's Share in Democracy," *Civil Engineering*, IX (November, 1939), 638.

45. *Ibid.*

46. A. B. Parsons, "Superorganizing Professional Engineers," *Mining and Metallurgy*, XXIV (September, 1943), 392.

47. Robert E. Doherty, "The Engineering Profession Tomorrow," *Journal of Engineering Education*, XXXV (September, 1944), 9.

48. Frederick S. Blackall, Jr., "ASME's Importance to Management," *Mechanical Engineering*, LXXV (September, 1953), 752.

49. *Who's Who in America*, 1930–1931, pp. 194–195.

50. Morris L. Cooke, "On the Organization of an Engineering Society," *Mechanical Engineering*, XLIII (May, 1921), 323, 325. See also Morris L. Cooke, *Professional Ethics and Social Change* (New York, 1946), 10.

51. Morris L. Cooke to Frederick W. Taylor, October 30, 1914, file "Cooke, July–December, 1914," Frederick William Taylor Collection, Stevens Institute of Technology, Hoboken, New Jersey.

52. Charles W. Baker to Edward W. Bemis, September 21, 1916, and Morris L. Cooke to Bemis, September 27, 1916, file "ASME Engineering Ethics–1916," box 168, Papers of Morris L. Cooke, Franklin D. Roosevelt Library, Hyde Park, New York.

53. "President Lee Answers Governor Pinchot," *Electrical Engineering*, L (March, 1931), 215.

54. Gregory M. Dexter, "An Appeal to Members" (letter to the editors), *Mechanical Engineering*, LXIV (October, 1942), 757.

55. W. H. Shockley, "The American Institute of Mining Engineers as Censor—A Protest," *Mining and Scientific Press*, CXIII (October 21, 1916), 589–590.

56. Robert B. Brinsmade, "Freedom of Discussion in the Institute" (letter to the editors), *Engineering and Mining Journal-Press*, CXIV (December 16, 1922), 1063.

57. "Code of Principles of Professional Conduct of the American Institute of Electrical Engineers," *Trans AIEE*, XXXI, pt. 2 (June–December, 1912), 2229, and "Report of Committee on Code of Ethics," *Trans ASME*, XXXVI (1914), 26. The ASME dropped this provision from its code of ethics in 1922.

58. Frank B. Copley, *Frederick W. Taylor*, 2 vols. (New York, 1923), II, 281.

59. Bernhard F. Jakobsen, *Ethics and the American Society of Civil Engineers* (Los Angeles, 1955), 1–13, and "A.S.C.E. Asked to Reinstate Two Los Angeles Engineers," *Engineering News-Record*, CXVII (October 8, 1936), 525.

60. Engineers Joint Council, *Directory of Engineering Societies and Related Organizations* (New York, 1963), 10–49. About a third of the national societies listed have company members; the amounts they provide are indicated in the individual societies' annual financial statements, which are usually published in their transactions.

61. Engineers Joint Council Subcommittee on Survey of Employer Practices, "Employer Practices Regarding Engineering Graduates," *Mechanical Engineering*, LXIX (April, 1947), 307.

62. Ernest K. Parks, "Membership in the A.I.M.E. is Purely Personal" (letter to the editors), *Mining and Metallurgy*, XXI (August, 1940), 387.

63. "Afraid to Ask?" *Mechanical Engineering*, LXX (January, 1948), 3.

64. *Ibid.*

65. Harry E. Harris, "Unification of Engineering Societies" (letter to the editors), *Mechanical Engineering*, LX (February, 1938), 177.

2 ▪ THE EVOLUTION
OF A PROFESSION

The balance between business and professionalism has been one of the most important forces in the formation and evolution of engineering societies in America. Most American engineering societies represent compromises between business and professionalism. A purely professional association might fail to win business support. Employer approval is needed, however, to permit employees to attend meetings or read papers. A wholly commercial organization, on the other hand, would tend to alienate creative professionals. And without their participation a technical society would fail of its fundamental purpose: to increase knowledge. Although the support of both professionals and businessmen is needed for a successful engineering society, the degree of influence accorded to each has varied widely; almost no two engineering societies are alike in this respect. However, the particular compromise adopted will have widespread influence on almost every aspect of the society's policies.

There are two fundamental ways of altering this balance: by defining the field to be covered and by professional standards of membership. Those who think of engineering as an independent profession have favored a comprehensive body that would represent and unite all engineers of whatever specialty. In contrast, engineers who think of themselves as businessmen have sometimes preferred technical societies built around a single industry and devoted to its interests. Strongly professional societies set high standards of membership that restrict effective power to initiates in a body of esoteric knowledge. Because such qualifications tend to exclude businessmen, management-oriented engineers have favored societies with lower membership requirements, which could bring together on an equal basis the businessmen, managers, and engineers affiliated with a given industry. As one engineer sympathetic to management put it, membership standards should be "inclusive and not too rigidly graded, granting to engineers who pass into administrative duties the full fellowship of the profession."[1]

Most engineers agree that their profession is a "vaguely bounded nucleus within a large body of technical workers."[2] Engineering societies mark off this professional nucleus by means of membership standards. At one time or another four different tests for full membership in engineering societies have been suggested: technical creativity, the ability to design, being in "responsible charge" of engineering work, and company or industrial affiliation. The crucial element in each is the degree to which businessmen—including managers—would be excluded from full membership. By the tests of technical creativity and design, almost all businessmen would be relegated to a class of associates; officeholding and power would be effectively restricted to professional engineers. The criterion of "responsible charge" would admit businessmen who performed some technical functions, but would exclude some others from full membership. A company or industrial standard would admit all businessmen who might be interested in full membership, but would effectively exclude most professional engineers. It is not uncommon for engineering societies to apply more than one of these tests.

Engineers closely allied with science have sometimes favored restricting full membership to those having done creative technical work. This is, in fact, the test that scientists have applied, though often informally, in determining who is a professional "insider." Such a standard would exclude from power all but a tiny minority of businessmen and managers. It would also exclude most of those who think of themselves as engineers. The problem is that the number of creative innovators is small. At the beginning of the twentieth century electrical technology was perhaps the most rapidly developing field of engineering; yet at that time only about 2 percent of the membership of the American Institute of Electrical Engineers was contributing papers in a given year.[3] A society restricted to the creative would have few full members. Nevertheless, creativity has been important, at least informally; the presidency of engineering societies is usually reserved for such men.

The test that most clearly distinguishes the professional engineer from all other groups is his ability to design engineering works. Placing the emphasis on the application of professional knowledge, rather than on its creation, distinguishes between the scientist and the engineer. Stressing the ability to design as well as to direct or construct engineering works draws a sharp line between professional engineers and managers and businessmen. Design has several drawbacks as a criterion of full membership, however. In actual fact, engineers perform a diversity of roles; they are not all designers. As a result, the requirement has been phrased as "qualified to design." This, in turn, is rather hypothetical and creates problems of administration. But however difficult of application, the intent is clear: to restrict full membership to professional engineers. It has, therefore, been adopted by the most strongly professional societies. Such organizations limit full membership to a professional elite; they exclude many persons who perform engineering work as well as most businessmen and managers.

A more inclusive test for full membership is that of being in "responsible charge" of engineering work. It is, in essence, a bureaucratic definition; it applies to engineers who do not have to refer their plans or drawings to a higher authority for technical review. Such a criterion of membership accepts the reality that the engineer in America is only partly differentiated from management. It is adapted to the heterogeneous tasks actually performed by engineers, who might be variously engaged in management, sales, production, testing, or any of a number of other functions. From the professional standpoint this criterion has a serious disadvantage: bureaucratic standing does not guarantee technical excellence. Despite this, "responsible charge" has been widely used as a professional qualification for membership in American engineering societies. The reason for its popularity is that it represents a compromise between professionalism and business. Those in charge of engineering work are in some sense professional engineers, but this test is broad enough to include most businessmen or managers who might be interested in membership. It is a common denominator between the engineer and the businessman.

Engineering societies that define their membership in terms of industrial affiliation represent a further dilution of professionalism. Such membership standards may be variously phrased. Some societies admit anyone interested in a given field of practice. If this field is defined narrowly enough, it will effectively limit membership to those in a given industry. This may be formalized by restricting membership to those "practically" engaged in the field. A further limitation is implied by a membership based on employment by a specific company or group of companies. The extreme of this tendency is reached when the technical society becomes a trade association, the litmus-paper test of this being the adoption of company rather than individual memberships. There is, however, an important borderland of societies that have both company and individual members. Societies with company members usually drop the term "engineering" from their titles; for example, the American Ceramic Society, the American Concrete Institute, and the American Waterworks Association.

Each of these four tests of full membership constitutes, in effect, a definition of the term "engineer." A membership qualification based on design implies that engineering is an independent profession, but one based on industrial affiliation suggests a basic solidarity between the engineer and the businessman. "Responsible charge" is a compromise which orients the engineer toward both his profession and his employer. "Creativity" tends to ally engineers with science rather than with either business or engineering. As recent sociological studies have indicated, these represent four possible orientations for the individual engineer in industry.[4] In practice, however, societies are seldom pure types. Most of them combine two or more of these tests of membership, in order to appeal to as many members as possible. By applying a diversity of qualifications, some perhaps informally, almost any desired combination of business and professionalism is possible. Consequently, engineering societies represent not four sharply defined types, but a continuous spectrum.

Engineering professionalism first appeared in America as an offshoot of the scientific variety. From 1829 to 1836, Alexander Dallas Bache utilized the Franklin Institute in an effort to advance professional science; a significant feature of these early activities was the attempt to gain patronage and prestige for science by linking it to practical activities, such as the investigation of steam-boiler explosions. Thus, when Bache and his scientific friends refashioned the institute's *Journal*, one of the three professional subdivisions they created was that of civil engineering. In this way, the institute became the first American center for the encouragement of a professional spirit in engineering.[5]

The first effort to form an engineering society in America occurred in February, 1839, when some forty engineers met at Baltimore. They appointed a committee to draft a constitution. This committee, logically enough, proposed to graft the new organization upon the Franklin Institute. The membership standards proposed would have pleased Bache: each member would be required to contribute at least one communication annually. British precedent was also important. A subcommittee quoted Thomas Telford, the first president of the British Institution of Civil Engineers, on the dangers of "too easy and promiscuous admission," which, he had warned, led to "unavoidable, and not infrequently incurable, inconveniences."[6]

The committee's proposals did not win the support of American civil engineers. There were several reasons for this failure. American engineers may not have relished such a close alliance with science; in any case, the membership standards were too high. Another reason may have been the desire of Pennsylvania engineers recently dismissed from their positions on state works to utilize the projected society for their own welfare. Sectional jealousies and indifference were, however, probably the most important factors. A counterproposal by Mr. Edward Miller, in 1840, was based on the idea of four independent regional societies.[7] In fact, the earliest engineering societies to endure were of precisely this character. The first was the Boston Society of Civil Engineers, founded in 1848, followed by the American Society of Civil Engineers, in 1852, which, despite its name, functioned initially as a local society for New York. Other early societies were the Western Society of Engineers, established at Chicago in 1869, and the Engineers Club of St. Louis, which appeared in 1868. Although local associations of engineers continued to multiply, their importance as carriers of professionalism was soon overshadowed by the emergence of national engineering societies.

Two engineering societies dominated the national scene in the 1870s; they represented something like thesis and antithesis in the dialectic between professionalism and business. The American Society of Civil Engineers claimed to represent all American engineers not in military service; it maintained high standards of membership that drew a sharp line between an elite of professional engineers and all others. But its ideals were challenged by the American Institute of Mining Engineers. The AIME did not attempt to represent the engineering profession. Its fundamental aim was service to the mining and metals industries. The AIME did not restrict its membership to professional engineers. The civil engineers were attempting to separate business and engineering, the mining engineers to merge the two. Fundamentally, the ASCE stood for the ideal of engineering as an independent profession, and the AIME embodied the notion that engineering was an integral part of business.

The ASCE was the first national engineering society to endure, and it set the pace for the professional development of engineering in America. Founded in 1852, the ASCE soon lapsed into a moribund condition, but it was revived in 1867. In its earliest days the ASCE had admitted all persons professionally interested in the advancement of engineering. Slightly more than two years after its revival, however, the ASCE created a new grade of associate, identical to its previous rank of member. Members were now required to have been in active professional practice

for five years and to be in charge of engineering work. In 1891, an even
sharper upgrading took place. A new rank of associate member was
created between member and associate, with requirements identical to
those previously specified for members. The full member now had to be
over thirty, have been in active practice of his profession for ten years,
and in "responsible charge" of engineering work for five years. Even
more significant was the requirement that a full member must be qualified to design as well as to direct engineering work.[8]

The ability to design was the crux of the ASCE's membership standards. Spokesmen for the ASCE maintained that the difference between
civil and other engineers was not that between coordinate branches of
engineering, but rather between professionals and nonprofessionals. As
one ASCE president put it in 1895,

> Any man who is thoroughly capable of understanding and
> handling a machine may be called a mechanical engineer, but
> only he who knows the principles behind that machine so thor-
> oughly that he would be able to design it or to adapt it to a new
> purpose . . . can be classed as a civil engineer.[9]

Thus, ability to design was rooted in esoteric knowledge. Without this
knowledge, the engineer was a workman rather than a professional.

The ASCE's high membership standards marked off an elite of full
professionals, established a graded hierarchy of professional excellence,
and protected the autonomy of the professional society. Realistically, the
chief threat to professional independence lay with the great railroad
corporations. A purely bureaucratic test for membership might have
given a decisive position in the society to railroad employees holding
managerial and supervisory positions. Although not without influence,
major business interests, such as the railroads, were balanced by the substantial number of consultants and government employees among
members. According to a tabulation in 1909, railroad employees, manufacturers, and contractors together constituted less than a third of all
members. Consultants and government employees totaled about one-
fifth each.[10] The consultants, in particular, constituted the cutting
edge of professionalism. They could be assumed to be lifelong practitioners. They were self-employed and presumably independent. High
membership standards enabled such professionals to dominate the
society.

After its revival in 1867, the ASCE evolved from a local to a national
society. In 1870, the society adopted the principle of holding its annual
meetings at various population centers across the country, which ena-

bled more of its members to attend and participate. Equally important were measures to decentralize the society's governmental machinery. In 1873, a letter-ballot was adopted to allow members distant from New York to vote in all elections. Five years later, the ASCE created a nominating committee to maintain geographical balance in the selection of officers. The society was divided into five geographical districts, and one member of the committee was selected from each district. These measures were not without success. By 1873, residents of the New York area constituted 30 percent of the society; by 1897, the proportion of nonresident members had climbed to 80 percent. [11]

The ASCE, however, paid a heavy price for its exclusive standards of membership. Its elitist tendencies alienated at least three important groups of engineers: the local groupings of civil engineers, the engineering specialists emerging in industry, and the younger engineers. Efforts by the ASCE to incorporate the local societies of civil engineers as branches failed because of the high membership standards of the national society. City engineers and others who were locally important did not relish the prospect of being relegated to inferior grades of membership in a national society. [12]

In the case of the engineers in industry, it was probably the ASCE's attempt to draw a sharp line between engineers and managers that was the source of difficulty. The typical mechanical engineer in the 1870s and 1880s was simply a plant superintendent or small manufacturer. One of the most widely respected mechanical engineers of this era was John Fritz. His early training had been as blacksmith and machinist; he worked his way up to the position of plant superintendent by self-training. He was active in the American Institute of Mining Engineers from its inception in 1871, but did not join the ASCE until 1893, a year after his retirement. [13]

The ASCE benefited the senior and successful men who qualified as full or associate members. Younger engineers who did not meet these standards found little to attract them to the ASCE; they were not even allowed to vote in society elections. Young civil engineers were to constitute the core of the protest organizations that arose to challenge the ASCE in the early years of the twentieth century.

The leaders of the ASCE do not appear to have regretted the failure of their society to become all inclusive. In fact, the society developed an elaborate set of procedures to maintain professional standards. Publications offer an important example. All papers submitted were first passed on by a committee. If accepted, the paper was distributed to the membership for comments; these monthly issues became the *Proceedings*.

After further scrutiny, the paper then appeared in the annual *Transactions*, along with the written comments of other members.[14] Committees were vested with the nomination of officers, the admission of new members, meetings, standardization, prizes, and similar functions. It was vital that this control machinery be kept in the hands of professionals. A more heterogeneous membership might allow nonprofessionals to take control and misuse these powers. The ASCE preferred to maintain its autonomy rather than to extend its influence.

The professional autonomy sought by the ASCE implied more than freedom from external control; its more profound meaning was the group's desire for moral independence. This required that practitioners look to colleagues for praise or disapproval, rather than to employers or the general public. The ASCE developed several mechanisms to enhance the importance of colleagual opinion. Informal contacts at society meetings were probably the most important. But they were supplemented by the practice of soliciting written comments on papers. Another way of expressing colleagual spirit was by awarding medals. In 1872, the ASCE established the Norman medal, the first such prize for American engineering. It was followed by several other prizes. Although never as significant as the major scientific prizes, these awards were designed to serve much the same purpose. The ASCE also encouraged colleagual spirit by participating in international professional gatherings. The ASCE was represented at the U.S. Centennial of 1876, the Paris Exposition of 1878, and the International Engineering Congress held in connection with the Columbian Exposition in Chicago in 1893.[15]

In relation to public policy questions, as with membership, the ASCE chose the safer course of guarding its independence rather than of attempting to expand its influence. It avoided antagonizing major interests. Although the railroads and other industrial groups were unable to control the society, they constituted powerful minorities that had to be appeased. Apart from the internal strife such groups could produce, there was the further danger that members might transfer their allegiance to rival societies. The formation of the American Railway Engineering Association in 1899 demonstrated how real this threat was. Thus, the ASCE largely ignored the delicate matter of railroad mismanagement. By confining its activities in sensitive areas such as standardization and safety to innocuous topics, the society avoided internal dissension and external pressures. But the result of the ASCE's prudence was that the society came close to abdicating its social and professional responsibilities.[16]

The ASCE's lack of aggressiveness extended to its relations with the army engineers. In 1872, the ASCE appointed a special committee to persuade the federal government to undertake tests of the strength of American iron and steel. The first effort along this line met with success, and Congress created a board of seven engineers: two from the army, two from the navy, and three from ASCE. But in 1878, Congress ended the board, and the testing machine that had been built was turned over to the Army Ordnance Department. The entire program seems to have become a casualty in the continuing hostilities between military and civilian engineers in America. Despite a congressional admonition in 1881 to "give attention to such programme of tests as may be submitted by the American Society of Civil Engineers," the army maintained that funds were insufficient to perform the tests desired by the civilian engineers.[17] The leaders of the ASCE avoided open conflict with the army and confined themselves to mild suggestions that a new testing machine be constructed for the use of civilian engineers.[18] They left it to engineers affiliated with local and state societies to continue the battle.

By the end of the nineteenth century, leaders of the ASCE viewed the progress of their society with a satisfaction bordering on complacency. The success of the ASCE rested on its very high professional standards of membership and the diversity of occupational roles open to civil engineers. Corporation employees were not in ascendency; nor was any single industry dominant. Presidents of the ASCE took a high degree of professional autonomy for granted. They depreciated the disagreements between civil engineers serving as expert witnesses before the courts and suggested, instead, that the professional engineer ought to serve the courts themselves rather than the litigants.[19] Such a role implied that engineers should be independent of specific commercial interests.

But the ASCE's professionalism was not without its drawbacks; the society did not serve the needs of the bulk of those who considered themselves professional engineers. It demanded a higher level of professionalism than the heterogeneous character of American engineering warranted; and, in consequence, rival societies sprang up that offered different balances between professionalism and business.

The first national society to dispute the ASCE's claim to represent all engineers was the American Institute of Mining Engineers, founded in 1871. In sharp contrast to the ASCE, the AIME showed little or no interest in professionalism. The AIME adopted a test of membership that was industrial rather than professional. It admitted anyone "practically engaged in mining, metallurgy, or metallurgical engineering."[20] Thus, the institute excluded the vast majority of professional engineers who were not affiliated with the mining and metals industries but admitted many nonprofessionals. Although the AIME created, on paper, the usual grades of membership, in practice virtually all of its members were full members. The purpose of the AIME was to serve the interests of the mining industry; professional development and welfare were not part of its program.[21]

From its inception in 1871 to 1912, the AIME was very much a one-man affair. The benevolent despot who ran the AIME was Rossiter W. Raymond. A man who knew Raymond well, Thomas A. Rickard, commented of his rule:

> As secretary of the Institute he performed divers duties; he invited written contributions and revised them before publication; he organized the meetings; he was the administrator. In course of time his ebullient personality so dominated the Institute that he was allowed a free hand to do as he thought fit. Presidents came and went; although nominally secretary, he exercised complete control. . . . Dr. Raymond managed its affairs, practically without let or hindrance.[22]

The institute was, in a sense, an outgrowth of Raymond's journalistic activities. He edited the *Engineering and Mining Journal,* in which early papers appeared concurrently with their presentation before the AIME.[23] Raymond's role as secretary was much like that of an editor; the AIME did not at first print discussions of its papers, as did all other American engineering societies. If a paper had Raymond's approval, that was sufficient.

Even though heterogeneous in membership, the AIME was not without a professional mission. The institute attracted a core of creative mining engineers and metallurgists who were engaged in something of a crusade to carry scientific methods to the mining and metals industries. Raymond himself was a highly trained professional. After graduation from Brooklyn Polytechnic Institute, he spent three years studying mining in Germany, notably at the famous Royal Mining Academy of Freiberg. Raymond's dominance eliminated any danger that the institute would be taken over by nonprofessionals.[24]

Raymond's professionalism, however, was limited. He thought of the engineer as a kind of businessman. In contrast, by the 1890s a number of younger mining engineers were arguing that engineering should be an independent profession. This difference appeared in, among other things, attempts to define the proper role of engineers in legal disputes. T. A. Rickard, one of the most vocal spokesmen for the younger generation, held the position favored by ASCE presidents: that engineers ought to serve the courts rather than the litigants.[25] Raymond, however, was an unreconstructed individualist with no inhibitions about the profit motive or commercialism. Always a robust partisan, he studied law and was admitted to the bar, the better to serve as advocate in mining cases.[26] If the AIME under Raymond was a professional society, it was one in which the balance between professionalism and business was heavily weighted on the side of business.

A substantial body of professionals became increasingly concerned over the absence of professional standards in the AIME. In 1885, President J. C. Bayles argued that the professional man should have a higher level of ethics than ordinary businessmen.[27] His suggestions for a code of ethics met with no response; Raymond successfully opposed all ethics codes for the AIME as long as he was its secretary.[28] An even more sensitive subject was the quality of the AIME's publications. Raymond strove heroically to maintain a high quality, often virtually rewriting the papers submitted. But the problem was inherent in the lack of membership standards, which brought many men into the society who lacked professional ability. President John Birkinbine was able to carry out a partial reform in 1893. Holding that it would be better to print "a small number of papers fully discussed" than a large number "offering controversial data," he instituted the practice of printing comments on papers.[29] This change did not satisfy the professionals, and their discontent was to lead to open rebellion against the Raymond regime early in the twentieth century.

Both the AIME and the ASCE represented balances between business and professionalism. But the AIME leaned so far toward business that it alienated professionals, and the ASCE was so exclusive that it failed to meet the needs of the engineering specialties emerging in industry. Consequently, in the 1880s two new societies appeared that attempted to combine the professionalism of the ASCE with the industrial service of the AIME. They were the American Society of Mechanical Engineers— or ASME—founded in 1880, and the American Institute of Electrical Engineers—or AIEE—founded in 1884. Both followed the formal structure of the ASCE; they were professional in spirit and set meaningful standards in membership and publications. But these standards were lower than those of the ASCE. Neither embodied the principle of professional unity, and both were closely allied with business interests. In their professional aspirations and the heterogeneity of their members, the societies of mechanical and electrical engineers reflected more accurately than their predecessors the character of the newer breed of industrial engineers.

The ASME was founded by three men whose careers illustrate the gradual transition of mechanical engineering from a craft-based art to a scientific profession. John E. Sweet exemplified the older generation. Without academic training, Sweet was variously pattern maker, draftsman, contractor, bridge builder, inventor, and manufacturer. In 1873, he was appointed "master mechanic" at Cornell's Sibley College, where he was director of the machine shop. In his later years Sweet was a small manufacturer and proprietor of a machine shop.[30]

Alexander L. Holley and Robert H. Thurston, in contrast, were prototypes of the newer generation of scientific engineers. Both earned engineering degrees at Brown, and both were concerned with importing to America a more rational, scientific technology. Holley witnessed the Bessemer process while on a visit to England in 1863. Thereafter, his career was devoted to the introduction of this method of steel making to American industry.[31] Thurston's emphasis was less on any particular innovation than on the scientific methods being applied to practical problems in Europe. He translated Sadi Carnot's classic treatise on thermodynamics into English. Perhaps his most characteristic innovation was to refashion the engineering curriculum of his day, giving a central place to the laboratory, where technological problems could be attacked by the methods of science.[32]

Sweet, Holley, and Thurston represented different traditions in mechanical engineering. Sweet followed the older pattern of artisan-engineer, and he found professional independence as the owner of a small business. Holley became a consultant, and he helped to bring into being a new age of giant corporations that made the entrepreneurial role obsolete in mechanical engineering. Thurston, as an educator, could devote himself to the future. He was the prophet of a new era in which science would provide the foundations for a new technology. All three were creative innovators. Their common dedication to the ideals of professionalism led to the foundation of the ASME.

The basic strategy of the ASME was to combine relatively low formal standards of membership with informal leadership by a professional elite. The member had to be "connected" with engineering; he did not have to be a professional engineer in active practice. The full member had to be in "responsible charge" of engineering works. Those who merely managed engineering works were theoretically not qualified for full membership; but in practice the ASME interpreted its rules loosely. Many plant superintendents, managers, contractors, and manufacturers who could not have met the society's standards if strictly applied were admitted to full membership.[33] It was symptomatic of the heterogeneous membership of the ASME that well over half of its members were owners or managers of businesses.[34] In effect, "responsible charge" became the basic membership requirement. As the secretary of the society, Frederick R. Hutton, put it: "The Society would have been a small one and of limited influence had its membership been restricted to the type of consulting or creative engineer alone. The factory engineer is more and more a manager of men. . . . The engineer must be what he is often called, a businessman."[35]

Despite its inclusive standards of membership, the ASME, at least until 1904, was firmly controlled by a minority of creative professionals. Such informal leadership was easy to establish because the society had been founded by them and they were the ones willing to devote their time to its government. The early leaders were able to keep power in the hands of professionals, in part because governmental policies and procedures were relatively fluid. The governing council and the president, with the secretary as executive officer, kept almost all powers in their own hands.[36] Guidance by such men was accepted because business-oriented engineers were principally consumers of technological knowledge; the creative minority produced new knowledge.

Informal professional leadership worked well only so long as the society was a small one. With increasing size, more formally organized, bureaucratic government became necessary. Local sections of members sprang up to challenge the power of headquarters. New specializations demanded, and got, a wide measure of autonomy. The secretary, Hutton, was well-meaning, but bumbling and inefficient, and the society began to run a deficit. When a proposal for a dues increase was rejected by the membership, sweeping governmental reforms had to be instituted. In 1904, the society set up a series of standing committees to conduct its affairs. The result of these changes was a wider degree of membership participation in society affairs.[37] But undermining informal professional leadership created the danger that nonprofessionals might gain control of the society. This threat was particularly great when the ASME's very active committee work in standardization and safety came to touch upon the private interests of members of the society.

How these private interests operated was illustrated in 1914 when the ASME drew up a standard code for steam boilers. The preliminary draft, drawn up by a committee headed by John A. Stevens, a distinguished engineer, aroused violent protests from members who were owners and executives of affected companies and who objected to rules interfering with their businesses. Led by Henry Hess, a member of the society's governing council, they demanded a reconstruction of the committee and the replacement of Stevens. A compromise was reached whereby a special advisory committee was created on which all of the major trade associations involved were represented, giving official recognition to the industrial interests concerned. Hearings were held and Stevens, who remained head of the committee, was able to reach agreement with all of the business interests concerned. After the code was amended to suit them, the trade associations banded together to lobby for the enactment of the boiler code by the states. In this instance the ASME was probably wise to recognize the limits of its autonomy, but the fact remained that it served more as an agent of industry than as an independent profession.[38]

Like the ASME, the AIEE represented a balance between business and professionalism. Where the ASME blended relatively low formal standards with informal professional leadership, the AIEE, by the end of the 1890s, combined very high formal standards with informal business leadership. This rather anomalous mixture was due to the fact that businessmen had provided the AIEE with its early guidance, but the rapid development of electrical engineering encouraged professionalism.

Unlike the ASME, the AIEE was not founded by professionals. The immediate stimulus for the formation of a new society was patriotic; a group of industrialists, inventors, and engineers thought there ought to be an American counterpart to the British Institution of Electrical Engineers to represent the United States at the International Electrical Exhibition to be held at Philadelphia in 1884. Nathaniel S. Keith, the man who issued the circulars calling for a new organization, was an inventor who appears to have been primarily interested in reforming the patent laws. He resigned as secretary of the AIEE after only a few months. Initial leadership for the fledgling organization was provided by executives in the telephone and telegraph industries. The chairman of the first meeting, Joseph P. Davis, was vice-president of a telephone company, and the first president of the AIEE was Norvin Green, the president of the Western Union Telegraph Company. Green was a doctor, legislator, and businessman; he was not a professional engineer. Business leadership became an institute tradition.[39]

The nature of electrical engineering made the AIEE one of the most professional of American engineering societies. The professionalization of engineering was everywhere associated with the shift from a craft to a scientific base for the underlying technology, but in probably no case was this transition so rapid as in the electrical field. Electrical engineering was not based on a centuries-old craft tradition; it was the child of science. It lent itself to treatment by the highly sophisticated methods of mathematical physics. The coming of alternating-current power and the growing complexities of communication systems created problems that the older inventor types were ill-equipped to solve. The AIEE flourished with the arrival of a new type of scientist-engineer who could master these technological difficulties.

Charles P. Steinmetz and Michael I. Pupin were leading examples of this new breed of electrical engineer. Both had received graduate training for the Ph.D. in Germany. Steinmetz worked out the mathematical laws governing the losses of efficiency in electrical apparatus due to alternating magnetism; his papers on "hysteresis" that he read before the AIEE in 1892 became classics.[40] Pupin gained fame as the discoverer of a method of improving long-distance telephony that rested on some recondite considerations of mathematical physics.[41] Pupin, Steinmetz, and a few kindred spirits set the pace for the AIEE. They not only gave important papers, but they frequently participated in discussions and commented on the papers of other members. They did much to make the AIEE a dynamic and exciting society in the 1890s.

The rise of professionalism in the AIEE was reflected in its membership standards. Keith's proposals had envisaged a composite society that would include engineers and businessmen.[42] The initial standards of membership were low and so vaguely worded that the secretary complained few members could understand them.[43] But the new professionalism was reflected in membership standards that, by the end of the 1890s, were almost as stringent as those of the ASCE. A full member had to be at least twenty-seven, a professional electrical engineer in active practice for at least five years, and in "responsible charge" for at least two years. Most important of all, the candidate had to be qualified to design as well as direct electrical engineering works. These requirements effectively excluded businessmen and managers who were not fully qualified engineers. Such men were relegated to the associate member class, along with younger engineers. Although able to vote, they were effectively excluded from power. While many businessmen were barred from full membership, scientists were welcomed. The close kinship of electrical engineering with science was recognized by an alternate qualification of having done original work of value to electrical science.[44]

The changing membership standards of the AIEE were accompanied by an important shift in orientation away from business and toward the profession. The early presidents, almost all of whom were important business executives, stressed the services of the institute to the electrical industry. In 1894, President Edwin J. Houston voiced the hope that the

AIEE in the coming decade "will be the acknowledged center of the industry and art it now so ably represents."[45] But by the early years of the twentieth century, presidents of the institute, such as Charles P. Steinmetz, tended to be important innovators, and the emphasis was on the advancement of the profession of electrical engineering, rather than on the electrical industry. Although the AIEE restricted its membership to those in the electrical field, it is significant that institute members took the lead in movements looking toward the unification of the engineering profession early in the twentieth century.

So long as the AIEE was small and its influence purely technical, businessmen seemed not to resent their exclusion from full membership. Business interests were well represented, since many of the early leaders of the electrical industry were engineers who had no difficulty in meeting the institute's membership standards. This situation changed as the AIEE grew and as its influence spilled over into areas of public policy. A membership drive in 1903 brought in 1,000 applications at a time when the total membership was only 1,630; 16 percent of these new applicants were managers whose duties were more executive than technical.[46] Many could not meet the qualifications for full membership; in the early 1900s nearly half of those applying for full membership were being rejected.[47]

Until 1912, the AIEE stood firm on the principle of restricting full membership to professional engineers. But the standard applied was not realistic. In 1905, only 14 percent of those who belonged to the AIEE were full members; the corresponding figure in the ASCE for the same year was 56 percent.[48] The chairman of the membership committee suggested that a special grade be established for business executives to spare them the humiliation of being placed in the same associate member class with younger engineers.[49] Such a compromise did not satisfy business leaders, since they would still be effectively excluded from power. The institute had embarked on a very ambitious program of committee work in standardization, safety, and public policy that bore directly on important business interests. The demands by businessmen for admission as full members were to create a major crisis in the AIEE in the period 1912 to 1913.

The four societies of civil, mechanical, mining, and electrical engineering were collectively termed the "founder societies"; they, like the ASCE earlier, were thought to comprise all engineering. Each of the founder societies attempted to maintain the integrity of its professional domain in order to preserve "the powerful group action which can be exercised by a large single organization."[50] To meet the needs of new groups of specialists, they created technical divisions. In most cases such divisions began as an extension of the traditional committee structure. Technical committees were created for special engineering fields; they were given wide latitude in organizing meetings, soliciting and passing upon papers, and spending money. In the course of time, these groups assumed virtually all the attributes of a separate society: they had their own governmental structure, elected their own officers, and, in large measure, managed their own affairs. In effect, the founder societies attempted to prevent secessions by converting themselves into federations of autonomous groups.[51]

The founder societies, however, were unable to prevent further societies from being formed in the fields claimed by them. New societies appeared almost yearly. Among them were the Society of Naval Architects and Marine Engineers, 1893; the American Society of Heating and Ventilating Engineers, 1894; the American Railway Engineering Association, 1899; the American Electrochemical Society, 1902; the Society of Automotive Engineers, 1904; the Illuminating Engineering Society, 1906; the American Institute of Chemical Engineers, 1908; and the Institute of Radio Engineers, 1912. Except for the chemical engineers, all of these were in fields claimed by at least one of the founder societies. The division of the engineering profession has continued down to the present; a count in 1963 listed 130 national engineering and allied societies.[52]

One reason for the formation of new societies was discontent with the particular balance between business and professionalism of the parent societies. Almost the only thing a separate society could do that could not be accomplished from within a professional division of an existing society was to alter this balance. The founder societies represented more than fields of practice; each had its own unique equilibrium between business and professionalism. Membership in the AIME implied not only an interest in mining and metallurgy, but an affinity for industrial service as well; conversely, membership in the ASCE carried with it a commitment to professionalism.

The histories of the Society of Automotive Engineers and the Institute of Radio Engineers are indicative of some of the forces working for the fragmentation of the engineering profession. The automotive engineers were in the field claimed by the ASME; the radio engineers in the domain of the AIEE. To some extent each of the newer societies was shaped by the nature of the field of technology that its members pursued. Automotive engineering, despite the spectacular growth of the industry, was technologically much more conservative and traditional. It was developed by practical men like Ford and Kettering, rather than by scientists like Steinmetz and Pupin. A weaker contact with science and less esoteric knowledge provided the basis for a closer affinity with business than with professionalism. Radio engineering was probably the most scientific of all technological areas and one of the most rapidly developing. The close contact with science, the esoteric quality of the knowledge, and the dynamism helped to make the radio engineers among the most professional in spirit.

Although the Society of Automotive Engineers adopted professional grades of membership similar to those of the ASME, it was oriented to industrial service from the start. Founded in 1904, the SAE drew its membership initially from the smaller automobile companies. These companies were in danger of being squeezed out of existence by the competition of the automotive giants. The SAE attempted to assist the small companies by means of technical standardization. Under the leadership of Howard E. Coffin, vice-president of the Hudson Motor Company and president of the SAE in 1910, this important industrial service was carried through with brilliant success. By 1918, the leaders of the large companies—Ford excepted—were anxious to join. The SAE had achieved its goal of an engineering society serving the automotive industry.[53]

But the aim of industrial service proved subversive to professionalism. The leaders of the SAE were eager to enroll "men of high standing in industry." But such individuals felt that the grades of membership created "class distinctions," since businessmen would be relegated to the associate grade of membership if they lacked qualifications as engineers. A group of members suggested that the SAE create two equal grades of Industrial Member and Technical Member in place of the older classes of Member and Associate. Since these two new grades would be identical in everything but name with the existing class of Member, the constitutional committee recommended that instead the society simply alter the requirements for full membership so that "distinguished service" to industry qualified a candidate equally with professional standing. This change converted the SAE into something approaching a trade association.[54] Further professional development was retarded. In 1965, for example, the National Society of Professional Engineers publically criticized the SAE for its failure to adopt a code of professional ethics.[55]

In the case of the Institute of Radio Engineers, similar origins led to different results. Early efforts at forming a separate society for radio engineers were directed at closer relations to industry. The germ of the

IRE was the Society of Wireless Telegraph Engineers, founded in 1907 by John S. Stone. Its membership was initially restricted to employees of the Stone Wireless Telegraph Company. Only gradually were its rolls opened to members of other companies. In May, 1908, Robert H. Marriott sent out a circular suggesting the formation of a more comprehensive society, the Wireless Institute, to be modeled after the AIEE. Marriott suggested a test of membership that was professional rather than industrial, namely that of "having done valuable, original work in Wireless." In May, 1912, these two organizations merged to form the IRE.[56]

A factor that enhanced the professionalism of the nascent IRE was the rise of business influence in the parent society, the AIEE. An important milestone in the shift of the AIEE from a professional to a business orientation was its adoption, in May, 1912, of new standards of membership that allowed business executives in the electrical industry to become full members, even though they might lack professional qualifications.[57] As a result of the lowering of standards by the AIEE, several prominent members transferred their activities to the IRE. The IRE required all members when elected to subscribe personally to the constitution, an apparent attempt to prevent the sort of palace revolution that had just taken place in the AIEE.[58] In 1914, President C. O. Mailloux of the AIEE denounced the secession of the radio engineers from the AIEE as without technical or economic justification. He was, of course, correct. The motivation for founding a new society was professional, not technical.[59]

The standards of membership adopted by the IRE were substantially the same as those of the AIEE prior to 1912. But informal factors were even more important in making the IRE an important center of professional spirit. Radio engineering was highly esoteric, and this tended to exclude nonprofessionals as much as did formal grades of membership. The institute attracted a small but dedicated group of creative professionals.[60] Initially directed to the needs of a single industry, the rapid proliferation of the electronics and radio industries soon provided the IRE with a highly diversified industrial base. Many of the IRE's members were trained in science, and the institute was as much oriented to science as to engineering. The pull of science was apparent in the decision not to add "American" to the IRE's name. Its members wanted the IRE to become an international society for all scientists and engineers engaged in radio research.[61]

The case of the IRE was exceptional; most technical societies based on a single industry have tended to be controlled by business. By restricting membership to persons affiliated with a single industry, the influence of that interest is automatically increased. This effect is magnified if, as was usually the case, professional standards were lowered to make it easier for businessmen to belong. Many such societies have adopted no professional grades of membership at all, but have admitted anyone practically interested in their subject matter. While rejecting professional criteria of membership, such societies have often accepted company members. About a third of the smaller national societies have company members; they constitute an important source of funds for such organizations.[62] At the extreme of this tendency are societies like the National Electric Light Association—later the Edison Electric Institute—that are wholly controlled by their company members. The NELA was, in everything but name, a trade association. It served as a lobbying and propaganda agency for the utilities.[63]

While business has provided much of the impetus toward division of the engineering profession, professionalism has encouraged a counterpressure working for integration and unity. In the limiting case of an all-inclusive society restricted to professional engineers, such as the ASCE once aspired to be, the power of any particular business interest would be minimized. By its sheer size such a society would have considerable power, which might be used to advance the interests of engineers as a group. For reasons such as these, unity has been the recurring dream of professional engineers. Proposals for unification appeared in the latter part of the nineteenth century, paralleling the formation of new societies. Although without immediate influence, they foreshadowed the much more serious efforts by engineers in the twentieth century to unify their profession.

One possible road to unity was through a confederation of local societies. On the invitation of the Cleveland Engineers Club, issued in May, 1880, four local societies met to form a national organization. Meeting in Chicago in December, a group of delegates organized the Association of Engineering Societies, at first composed of groups from Chicago, Boston, Cleveland, and St. Louis. Later enlarged, this association lasted for thiry-five years. Its chief function was to publish a technical journal in order to give wider circulation to papers presented before member societies. The group had larger aims, however, such as encouraging the formation of a local engineering society in each center of population able to support one. The association was designed to provide a model for a decentralized unity organization of engineers, to offset the rising power of the national societies. Despite some vague talk of advancing common interests, little was accomplished, and the association was soon overshadowed by the founder societies. Perhaps its chief professional service was to publish an index of engineering periodicals. [64]

Within five years this early indication of group solidarity by the local societies was followed by evidence of a concern for the national government's mishandling of public works. Civilian engineers objected to the log-rolling that led to the selection of projects according to political rather than by engineering criteria; they also resented the dominant position of the army engineers in national public works. An attempt at reform came on the initiative of Cleveland engineers who called a meeting for December, 1885, to discuss national public works. Attended by representatives of ten local and state societies, the meeting generated enough interest to lead to the formation, at a second conference the following year, of the Council of Engineering Societies on National Public Works. Composed of twenty-three local and state engineering societies, it was led by several well-known engineers. Its aim was to secure reforms in the way the national government handled its public works. To this end, the council drew up a bill and was able to get it introduced into Congress; little else seems to have been accomplished, and the council soon went out of existence. [65]

In 1886, a prominent engineer, William Kent, proposed that public-works reform be merely one part of a more drastic reorganization of the engineering profession. In a speech before the American Association for the Advancement of Science, Kent suggested unification at the top through an Academy of Engineering. It was to be an "aristocracy based on intellect and achievement," in which engineers of all types would overcome other loyalties for the larger glory of their profession. The academy, as Kent envisaged it, would lead the profession by publishing technical papers, conferring medals, and sponsoring research. Although the academy was to be a private body free of any outside control, Kent optimistically thought that Congress might be persuaded to vest it with responsibility for both the planning and construction of all government public works. In this sense the academy would function as a governmental agency, but one controlled by the profession rather than by the representatives of the people. [66]

A more mundane proposal for unity was made by H. F. J. Porter, before the ASME in 1892. He suggested that the engineering congress to be held in association with the Columbian Exposition in 1893 be authorized to create a permanent committee that would function as a national agency for the engineering profession. Porter's emphasis was on bread-and-butter issues; the committee would standardize engineering education and license professional engineers. He also suggested measures looking to the regulation of competition among engineers and the establishment of a code of ethics. In effect, Porter wanted to apply to engineering the formula so successfully employed by doctors and lawyers for their professions.[67]

The spirit of professional unity in the founder societies was inhibited by the aloofness of the ASCE. The civil engineers maintained that the ASCE was the proper unity organization for all nonmilitary engineers, and it refused to consider cooperation with other societies. However, tentative efforts to secure at least a first step in this direction were made by members of the ASME and the AIEE. Two suggestions appeared in the 1880s: for a common library and for a joint headquarters building. In 1889, the ASME appointed a committee on a joint building for engineering societies, but apparently no progress was made.[68] Proposals for library cooperation met with no success either, as far as the ASCE was concerned, although the ASME and AIEE did cooperate to a limited extent. These meager returns fail to indicate the amount of interest and enthusiasm that was aroused by the prospect of closer cooperation by the founder societies. Much discussion took place informally, particularly at the Engineers Club, a private organization of New York engineers.[69] These discussions were to bear fruit in 1903 when Andrew Carnegie gave the founder societies funds for a common headquarters building.

Most of the early proposals for unification of the engineering profession were Utopian; but the desire for unity was growing. These early suggestions pointed to three possible methods of unity, all of which were to be explored in the twentieth century: a confederation of local societies, a separate unity organization, and a federation based on the founder societies. What was needed to give substance to these hopes was a further development of professional spirit. This condition was to be amply fulfilled between 1900 and 1920. In this process, professionalism was to be transformed from a matter of private loyalty to a crusading social movement.

NOTES

1. William E. Wickenden, "Professional Status of the Engineer," *Civil Engineering*, I (October, 1930), 25.

2. William E. Wickenden, "Engineering Education Needs a 'Second Mile,' " *Electrical Engineering*, LIV (May, 1935), 472.

3. Charles F. Scott, "Proposed Developments of the Institute," *Trans AIEE*, XX (1902), 9.

4. William Kornhauser, *Scientists in Industry, Conflict and Accommodation* (Berkeley and Los Angeles, 1962), 118–121. For a discussion of the literature on orientation, see Eugene S. Uyeki, "Behavior and Self-Identity of Federal Scientist-Administrators," in H. D. Lerner, ed., *Proceedings of the Conference on Research Program Effectiveness* (New York, 1966), 497–498.

5. Joseph B. Sinclair, " 'Science with Practice; Practice with Science': A History of the Franklin Institute, 1824–1837" (Ph.D. dissertation, Case Institute of Technology, 1966), 102–120, 142–152, 163–183, 198–210.

6. Charles Warren Hunt, *Historical Sketch of the American Society of Civil Engineers* (New York, 1897), 12, and "Civil Engineering," *The Journal of the Franklin Institute*, XXIII, n.s. (March, 1839), 160–167.

7. Daniel Hovey Calhoun, *The American Civil Engineer, Origins and Conflict* (Cambridge, 1960), 184–185.

8. Hunt, *Historical Sketch*, 17, 43–44, 77. See also John W. Lieb, Jr., "The Organization and Administration of National Engineering Societies," *Trans AIEE*, XXIV (1905), 286–287, and Raymond Harland Merritt, "Engineering and American Culture, 1850–1875" (Ph.D. thesis, University of Minnesota, 1968), 153–157, *passim*.

9. George S. Morison, "Address at the Annual Convention," *Trans ASCE*, XXXIII (January–June, 1895), 472.

10. Onward Bates, "Address at the 41st Annual Convention," *Trans ASCE*, XLIV (September, 1909), 570.

11. Hunt, *Historical Sketch*, 48, 50, 73–75.

12. *Ibid.*, 74, 76, and A. B. Stickney, "A National Spokesman for Engineers," *The Journal of Engineering Education*, XXXVI (April, 1946), 516.

13. John Fritz, *The Autobiography of John Fritz* (New York, 1912), 215.

14. Hunt, *Historical Sketch*, 66–68.

15. *Ibid.*, 49, 57, 62–65.

16. *Ibid.*, 83–84, provides a summary of committee work. Some civil engineers, notably John B. Jervis and Albert Fink, sought to reform railroad management by extending to it some of the rational, quantitative methods characteristic of engineering (Merritt, "Engineering and American Culture," 96–124).

17. Hunt, *Historical Sketch*, 83.

18. Henry Flad, "Address at the Annual Convention," *Trans ASCE*, XV (1886), 514.

19. William P. Shinn, "Address at the Annual Convention," *Trans ASCE*, XXII (June, 1890), 383.

20. "Rules," *Trans AIME*, I (1871–1873), xvii.

21. In 1905 there were 3,680 who belonged to the AIME, all but 190 of whom were full members (John W. Lieb, Jr., "The Organization and Administration of National Engineering Societies," *Trans AIEE*, XXIV [1905], 285, 287). It is significant that when the AIME was reorganized in the 1950s, the Society of Mining Engineers Division, the core of the old AIME, had 11,948 company members, thus formally ratifying what had been *de facto* the case all along: that the AIME was something intermediate between a professional society and a trade association (Engineers Joint Council, *Directory of Engineering Societies and Related Organizations* [New York, 1963], 39).

22. Thomas A. Rickard, *Rossiter Worthington Raymond; a Memorial* (New York, 1920), 9.

23. "Preface," *Trans AIME*, I (1871–1873), iii.

24. Presidents of the AIME were drawn exclusively from this professional core. For a list of presidents, with short biographical sketches, see A. B. Parsons, ed., *Seventy-five Years of Progress in the Mineral Industry, 1871–1946* (New York, 1947), 495–511.

25. T. A. Rickard, *Retrospect, An Autobiography* (New York, 1937), 66–67.

26. *Dictionary of American Biography*, s.v. "Raymond, Rossiter Worthington." For an example of Raymond's philosophy, see his "The Conservation of National Resources by Legislation," *Bulletin of the American Institute of Mining Engineers* (May, 1909), appendix, 20–36.

27. J. C. Bayles, "Professional Ethics," *Trans AIME*, XIV (1885–1886), 609.

28. R. W. Raymond, "Professional Ethics," *Bulletin of the American Institute of Mining Engineers* (January, 1910), 43–50.

29. John Birkinbine, "The Development of Technical Societies," *Trans AIME*, XXI (1892–1893), 971.

30. Albert W. Smith, *John Edson Sweet* (New York, 1925), 25–41, 65–69; *D.A.B.*, s.v. "Sweet, John Edson."

31. *D.A.B.*, s.v. "Holley, Alexander L."

32. William F. Durand, *Robert Henry Thurston* (New York, 1929), 62–71, 149–160, 191–200. Despite Thurston's leadership, apprenticeship training remained for many years an important source not only of mechanical engineers but of professionalism as well (see Monte A. Calvert, *The Mechanical Engineer in America, 1830–1910* [Baltimore, 1967], 43–62, *passim*).

33. The 1904 constitution required that "a Member must have been so connected with engineering as to be competent, as a designer or as a constructor, to take responsible charge of work in his branch of engineering." The standard applied from 1884 to 1903 was substantially the same except that "in the opinion of the Council" was placed before "competent," thus opening the door to administrative discretion in applying the requirements ("Rules of the American Society of Mechanical Engineers," *Trans ASME*, VII [1885–1886], xxv–xxvi, and "American Society of Mechanical Engineers, Constitution," *Trans ASME*, XXV [1904], xii). On the laxness of actual administration, see Frederick R. Hutton, *A History of the American Society of Mechanical Engineers from 1880 to 1915* (New York, 1915), 27.

34. Frederick R. Hutton, "The Mechanical Engineer and the Function of the Engineering Society," *Trans ASME*, XXIX (1907), 643.

35. Hutton, *History of the ASME*, 26.

36. *Ibid.*, 17–18, 23, 70, 78–133.

37. *Ibid.*, 70–74, 108–110, 290–294.

38. John A. Stevens, "Work of the Boiler Code Committee," *Journal ASME*, XXXIX (October, 1917), 855–857, and Arthur M. Green, Jr., "The ASME Boiler Code," *Mechanical Engineering*, LXXIV (August, 1952), 641–642.

39. "The American Institute of Electrical Engineers," *Trans AIEE*, I (1884), 1–9; *D.A.B.*, s.v. "Green, Norvin."

40. Charles P. Steinmetz, "On the Law of Hysteresis," *Trans AIEE*, IX (January–December, 1892), 3–64, 621–758. For a recent assessment of Steinmetz, see John Adams Miller, *Modern Jupiter, the Story of Charles Proteus Steinmetz* (New York, 1958), 48–51.

41. Michael Pupin, *From Immigrant to Inventor* (New York, 1923), 330–340. James E. Brittain, "The Introduction of the Loading Coil: George A. Campbell and Michael I. Pupin," *Technology and Culture*, XI (January, 1970), 36–57, shows that Pupin's contributions in this matter were less important than those of Oliver Heaviside and George Campbell.

42. Nathaniel S. Keith, "The American Institute of Electrical Engineers," *Trans AIEE*, I (1884), 1.

43. "AIEE Annual Meeting," *Trans AIEE*, II (1885), 8, and "Appendix, American Institute of Electrical Engineers Rules," *Trans AIEE*, I (1884), 1. Full members were supposed to be "professional electrical engineers," but the meaning of this term was not further specified.

44. John W. Lieb, Jr., "The Organization and Administration of National Engineering Societies," *Trans AIEE*, XXIV (1905), 287, and American Institute of Electrical Engineers, *Yearbook* (1912), 17, 23–24. Associates could vote and hold offices, excepting only those of president and vice-president. Exclusion from the vice-presidency had the effect of limiting their power on the society's governing council; but the lack of prestige of the associate membership was a far more effective means of limiting the power of businessmen.

45. Edwin J. Houston, "A Review of the Progress of the American Institute of Electrical Engineers," *Trans AIEE*, XI (1894), 282.

46. Charles F. Scott, "President's Address," *Trans AIEE*, XXII (1903), 7.

47. "Annual Report of the Board of Examiners," *Trans AIEE*, XXIV (1905), 1138.

48. John W. Lieb, Jr., "The Organization and Administration of National Engineering Societies," *Trans AIEE*, XXIV (1905), 285. Full members constituted 69 percent of the ASME and 94 percent of the AIME.

49. "Annual Report of the Board of Examiners," *Trans AIEE*, XXIV (1905), 1140.

50. "The ASME Fiftieth Annual Meeting," *Mechanical Engineering*, LII (January, 1930), 89.

51. *Ibid.* See also Hutton, *History of ASME*, 293–294.

52. Engineers Joint Council, *Directory of Engineering Societies*, 10–45.

53. George V. Thompson, "Intercompany Technical Standardization in the Early American Automobile Industry," *The Journal of Economic History*, XIV (Winter, 1954), 1–20.

54. "Report on Constitutional Amendments," *The Journal of the Society of Automotive Engineers*, II (January, 1918), 5–6.

55. "Engineers Debate Auto Safety Role," *New York Times*, April 25, 1966, p. 47.

56. "The Genesis of IRE," *Proceedings of the IRE*, XL (May, 1952), 517, and George H. Clark, *The Life of John Stone Stone* (San Diego, 1946), 133–137.

57. American Institute of Electrical Engineers, *Year Book* (1913), 27, and "Report of Board of Directors," *Trans AIEE*, XXXII, pt. 2 (May–December, 1913), 2168. The amendment admitted as full member "an executive of an electrical enterprise of large scope," whose "standing" was equivalent to that required of members who qualified as professional electrical engineers.

58. IRE *Year Book* (1914), 15. See also Laurens E. Whittemore, "The Institute of Radio Engineers—Forty–Five Years of Service," *Proceedings of the IRE*, XLV (May, 1957), 603.

59. C. O. Mailloux, "The Evolution of the Institute and of Its Members," *Trans AIEE*, XXXIII, pt. 1 (January–June, 1914), 823. Mailloux did not mention the IRE by name, but he specifically exempted the only other societies involved, the American Electrochemical Society and the Illuminating Engineering Society.

60. According to the original constitution, full members were to be persons "of good professional standing" with five years experience. In 1915, the IRE adopted a new set of membership standards providing for five grades of membership: honorary member, fellow, member, associate, and junior. The grades fellow and member corresponded roughly to those of member and associate in the AIEE prior to 1912. A fellow had to be "qualified to design and to take responsible charge" of radio work, or be a person "who has done notable original work in radio." Businessmen were admitted as members, but with a proviso that they be "qualified to take responsible charge of the broader features of radio engineering," thus admitting those in technical management, but excluding others. Members, like associates in the AIEE prior to 1912, were not eligible for the presidency or vice-presidency (IRE *Year Book* [1914], 7–8, 13; IRE *Year Book* [1916], 25–27). On the informal character of radio engineering, see Robert H. Marriott, "United States Radio Development," *Proceedings of the IRE*, V (June, 1917), 195.

61. In considerable measure the institute was successful in this; it attracted a substantial number of foreign members. Beginning in the 1930s, it became customary for the vice-president to be from a country other than the United States. In 1957, neither the president nor vice-president was a resident of this country (Laurens E. Whittemore, "The Institute of Radio Engineers—Forty-Five Years of Service," *Proceedings of the IRE*, XLV [May, 1957], 601).

62. Engineers Joint Council, *Directory of Engineering Societies*, 10–45. Thirty-nine of the societies listed have company members; not all the societies listed, however, are professional organizations of engineers.

63. Morris L. Cooke, "Nation-Wide Organization of the Utility Industries," in Morris L. Cooke, ed., *Public Utility Regulation* (New York, 1924), 296–298.

64. Gardner S. Williams, "Engineering Cooperation Outside the National Societies," *Bulletin of the Federated American Engineering Societies*, II (April, 1923), 5–6.

65. *Ibid.*, 6, and Frederick H. Newell, "An Engineering Council Now Almost Forgotten," *Engineering News-Record*, LXXX (April 25, 1918), 806–808.

66. William Kent, "Proposal for an American Academy of Engineering," *Van Nostrand's Engineering Magazine*, XXXV (1886), 277–280.

67. H. F. J. Porter, "How Can the Present Status of the Engineering Profession be Improved," *Trans ASME*, XIV (1892–1893), 487–497.

68. Hutton, *History of ASME*, 176.

69. *Ibid.*, 179–180, 182–183, and Hunt, *Historical Sketch*, 61.

3 • THE IDEOLOGY
OF ENGINEERING

Engineers, as a rule, are not and do not pretend to be philosophers in the sense of building up consistent systems of thought following logically from certain premises. If anything, they pride themselves on being hard-headed practical men concerned only with facts, disdaining mere speculation or opinion. In practice, however, engineers do make many assumptions about the nature of the universe, of man, and of society. These premises, in turn, influence their conceptions of their role in society and responsibilities toward it. Each idea is usually treated as if it were quite independent of any metaphysical system; that is, as a matter of self-evident truth scarcely open to question. Being implicit, or at least accepted uncritically, these ideas often lead to highly contradictory results.

Although it is incorrect to regard engineers' thoughts as forming a single, consistent body of doctrine, there are wide areas of agreement and many uniformities. Their most common ideas are, in fact, logically related. They can be arranged in sequences, which form systems of thought following from major premises. So ordered, the engineers' thoughts contain not one but two independent systems that are interwoven with one another, but that are not logically commensurable. One is materialistic, emphasizing scientific laws and the material environment. The second is idealistic, stressing *a priori* ethical imperatives and moralism. Of the two the scientific is the more distinctive; the idealistic is simply a variant of the thinking of American businessmen. In short, tensions between business and science have left as deep an imprint on engineers' thinking as on other areas of professional experience.

Engineers have obscured the tensions in their thought by using the term "law" ambiguously. Both the scientific and moralistic systems of thought may be considered to rest on a single premise: that the world is governed by immutable laws which men may know. Thus, George F. Swain, a prominent civil engineer, believed that "we live in a world of natural law. For everything that is, there is a cause, not only in the physical world, but in the mental and moral world; and this cause will always produce its effect."[1] This reconciliation is, however, only verbal. Natural laws cannot be violated, but moral laws can. Scientific laws are materialistic, moral laws idealistic. In the moralistic view man is a free agent, but in the scientific perspective man and society are assumed to be ruled by laws similar to those of mechanics or thermodynamics. Each system of thought implies different remedies for social ills: in the one case by a reassertion of correct moral principles, in the other by under-standing and controlling the environment.

The divided mind of the engineer reflects social as well as logical tensions. Each of the systems of ideas of the engineer carries with it a particular group loyalty and identification. The moralist perspective relates the engineer to business and, broadly, to middle-class values. Its component ideas, such as individualism, constitute the ideological underpinnings of American capitalism. As an identification, this thinking does not distinguish the engineer from the businessman. In contrast, the scientific frame of reference has, as its very foundation, emphasis on the esoteric knowledge that differentiates the engineer from the business-man, and it readily leads to doctrines of professionalism, such as the need for autonomy and colleagual control of engineering. Thus, one set of ideas identifies the engineer as a particular kind of businessman, and the other marks him off as a unique and superior occupational type. This division in thinking does not divide engineers into two distinct groups of businessmen and engineers; on the contrary, elements of both are almost always blended in the thinking of individual engineers.

The ideological rift in the thinking of engineers had its origins in the period from 1895 to 1920; it followed the movement of the engineer into industry. Prior to that time the ideas of engineers apparently did not differ fundamentally from those of the rest of the population. Although tensions between professional independence and business loyalty were present from the earliest appearance of the civil engineer in America, these conflicts were not ideological in the sense that they did not extend from specific issues to metaphysical systems or general theo-ries of politics. As the locus of engineering practice shifted to giant industrial corporations after about 1880, professional values were chal-lenged by larger and more rigorously authoritarian bureaucratic struc-tures. It was perhaps as a defensive reaction to this threat that engineers incorporated their values within an ideology. This ideology provided powerful new defenses for professional values by linking them to basic principles of universal significance. It went further: by demonstrating that they were necessary to social progress and the good of humanity, it made their defense morally binding on engineers.

Leading engineers in the 1880s and 1890s were deeply influenced by Herbert Spencer's social Darwinism; it contained many of the germs of an ideology of engineering. Spencer himself was an engineer. Engineers regarded him not only as a colleague but as an example of a new professional role: that of lawgiver and philosopher. Spencer was taken by Robert Thurston, the first president of the ASME, as having vastly widened the legitimate concern of the engineering profession to include general questions of politics and economics. Such a larger scope was professional, since Spencer had supposedly proved that man and society were governed by scientific laws of the same character as those used by engineers. Thurston's suggestion that the ASME become an active force in politics thus merely reiterated the traditional engineering role of applying known natural laws to meet practical needs. Thurston also found in Spencer's writings warrant for the identity of "scientifically correct conduct" and "righteous conduct." This implied that truth and justice were one and the same, and that engineering solutions to social problems would also be just solutions. [2]

Although engineers were greatly influenced by Spencer, they did not accept the deterministic implications of his teachings, certainly not as implying any limits on their own freedom of action. In discussing social evolution, engineers emphasized the importance of technological innovation. But to engineers inventions were free creations of the human spirit, not the determinate consequences of natural law. Thus, in 1892, Charles H. Loring argued that the steam engine had made it possible for man to abolish slavery and had opened up vistas for future progress undreamed of by the ancients. [3] Similarly, engineers gloried in competition and survival of the fittest, not because these principles in any way limited them, but because they were guarantees of their own success. Their scientific knowledge gave them a competitive advantage that insured their eventual triumph. In their stress on the creative role of human intelligence in evolution, engineers not only undermined determinism, but converted Spencerianism to something vaguely resembling the reform Darwinism of Lester Ward.

Professional leaders found it difficult to reconcile Spencer's philosophy with professional values. Although Spencerianism doubtlessly allowed engineers to relate their work in large corporations to the good of humanity, it offered no obvious support for those things that engineers thought good for themselves. It failed to provide engineers with a distinct identity and social role; it did not give ideological justification for professional values. Thurston argued that a policy of *laissez faire* by engineering societies in professional matters would be the best means for advancing the engineer's own interests in the long run.[4] But this offered small consolation to those who felt that deeply cherished values were in danger in the short run. Thurston thought of the engineer as a businessman. His proposals for an enlarged political role for the ASME, had they been adopted, would have diminished the profession's sense of constituting a separate group. He envisaged engineers participating in a movement of businessmen to influence legislation. The specific proposals suggested by him included upholding the protective tariff and crushing strikes; he did not foresee an independent or distinctive role for engineers.[5]

Professional values made sense only by reference to a distinct group with a proprietary attitude toward a body of knowledge and a field of practice. Thurston and other engineers influenced by Spencer denied their profession precisely this sort of monopolistic claim over technology. They saw it, perhaps realistically, as a joint product of men of many occupations: engineers, businessmen, scientists, and workers. Loring, in his discussion of the steam engine in history, did not suggest that it was the unique contribution of any single group.[6] A. L. Holley, along with Thurston one of the founders of the ASME, thought that mechanical engineering lay at the root of all technology and, by implication, of all material progress. But he did not think the engineer had an exclusive claim on this technology. On the contrary, he thought one of the ASME's most useful functions would be that of bringing together businessmen and engineers associated with mechanical engineering.[7]

Engineers created an ideology of their own by grafting professional values onto Spencerian metaphysics. By asserting that all technology was the work of engineers, they defined their social role. By holding that technology was applied science, they laid claim to a sophisticated body of knowledge. From these fundamental postulates of esoteric knowledge and social service, all of the values of professionalism could readily be derived. Although engineers thought of themselves as functioning in this role in an essentially Spencerian universe, they showed little interest in the remoter philosophical implications of their assumptions. Rather, they used them ideologically to defend their group and its interests.

Since professional values were built into its very structure, the engineers' ideology constituted a philosophy of professionalism. Through it engineers could find an identity as professional men distinct from the business community. By its means engineers could seek a measure of freedom from the rigors of bureaucratic control. This philosophy involved a shift in emphasis from technology to technologists. Where Spencer had started with laws of nature, engineers would shift the emphasis to the men who manipulated these laws for social purposes. Thus, the previous doctrine of the creative role of technology in social evolution would lead to the idea that the engineer had the future in his hands. Where Spencer had reduced political policy to the dogma of *laissez faire*, engineers would look forward to an indeterminate and flexible engineering of society based on their own ability to apply scientific knowledge for utilitarian ends. The future society would be what the engineering profession willed it to be.

Engineering societies offered an organizational framework through which engineers might hope to achieve their professional aspirations. But first their members had to be indoctrinated with these new ideas. The annual addresses of the presidents provided an early means for articulating an ideology of engineering. These speeches, with a few notable exceptions, had been traditionally devoted to technology; the constitution of the ASCE specified that the presidential address should consist of a review of the year's progress in engineering. In 1895, President George S. Morison broke with precedent; his speech was devoted to "the true meaning and position of the profession."[8] It was the first fairly complete statement of the new ideology of engineering. Morison's example was gradually emulated by succeeding presidents of the ASCE and, after 1900, by those of the AIEE. Spokesmen for the ASME and the AIME fell in line only a decade later; but after approximately 1910, the new ideas gained momentum. By 1920, the philosophy of professionalism had become something of an obsession with engineers.

Three themes served at once to express and to encourage this new ideology. In speeches delivered in the period 1895 to 1920 before major engineering societies, presidents and others in the vanguard of professionalism portrayed the engineer in glowing terms. They saw him as the agent of all technological change, and hence as a vital force for human progress and enlightenment. Secondly, such men drew an image of the engineer as a logical thinker free of bias and thus suited for the role of social leader and arbiter between classes. Finally, these speeches indicated that the engineer had a special social responsibility to protect progress and to insure that technological change led to human benefit. The idea that a particular group has a specific social role, a unique set of characteristics, and a particular mission to perform implies that the individuals composing the group identify themselves with the group. Thus, engineers had to think of themselves as engineers before they could think of their responsibilities as engineers. Those engineers who thought of themselves as businessmen might be concerned, like Thurston, with social responsibilities; but, like him, they would define these responsibilities in relation to the business community, not to the engineering profession.

The cement binding the engineer to his profession was scientific knowledge. All of the themes leading toward a closer identification of the engineer with his profession rested on the assumption that the engineer was an applied scientist. It was the cumulative character of scientific knowledge that gave weight to engineers' claims to be the agents of progress and enlightenment. Similarly, the self-image involved transferring to the group attributes of science such as logic and impartiality. Engineers thought that they could discharge their social responsibilities by applying their expertise to politics. The supposed universality of scientific laws and methods gave warrant to such an extension of their professional domain. All of these themes thus rested on the scientific revolution taking place in technology. Together they had the effect of marking off the engineer from other groups and of asserting his superiority to them.

The foundation of the engineers' ideology was the assumption that their group had a unique and vital role to play in social progress. An archetypical example of what was to become a very common theme of speeches before engineering societies was the precedent-breaking address of George S. Morison before the ASCE in 1895. Morison maintained that the evolution of civilization was based on inventions, such as fire, pottery, the manufacture of iron, and the production of power. Each of these developments determined an epoch in human history; mental and moral improvements were made possible by these changes in the material conditions of life. Morison differed sharply from Thurston, Loring, Holley, and others in claiming sole credit for all of these material advances for the engineering profession. In effect he defined all technological innovators as engineers. Thus, one of the youngest professions was given at once an ancient lineage and an exclusive claim to all future technological progress. Morison also provided the engineer with a mission:

> We are the priests of material development, of the work which enables other men to enjoy the fruits of the great sources of

power in Nature, and of the power of mind over matter. We are priests of the new epoch, without superstitions.[9]

The engineer, to Morison, was not a businessman; he was a superior being destined to supplant the businessman. "Accurate engineering knowledge must succeed commercial guesses," Morison maintained. "Corporations, both public and private, must be handled as if they were machines."[10]

The engineer emerged in the speeches before major engineering societies as not only the source of material progress but of enlightenment as well. Charles P. Steinmetz thought that "empirical science" had performed an important service in defeating and discrediting "metaphysics." He asserted that "at the entrance of the twentieth century metaphysics has practically ceased to be considered, and empirical science is universally acknowledged as the source of all human progress."[11] Not all advocates of the enlightening role of engineering were as positivistic as Steinmetz. Gano Dunn, another prominent electrical engineer, thought engineering had no influence on art, religion, and other areas of the human spirit. But he maintained that the engineer's way of thinking was destined to become "the way almost everybody thinks" and that much of the engineer's special knowledge was eventually to become general knowledge. The source of confidence in engineering's intellectual mission for both Steinmetz and Dunn was their identification of engineering with science. To Dunn the engineering method was simply "the applying to practical and utilitarian ends the principles and reasoning of science."[12] By this line of argument all the material and most of the intellectual benefits of science to civilization should be claimed by the engineering profession.

Engineers at the turn of the century repeated the idea that civilization was entering a "new epoch" and that engineering was building a "new civilization." These visions of the future verged on the Utopian at times; but they rested on the assurance that applied science carried within it the potentiality of unparalleled material abundance and leisure. The engineer's work appeared to one professional spokesman as the "herald that brings joy to the multitudes . . . it is their redeemer from despairing drudgery and burdensome labor."[13] Material achievements would benefit humanity and advance civilization. Technological progress was leading toward universal peace and the brotherhood of man. One mechanical engineer argued that his profession held a commission from God to subdue the earth. The technical marvels of the future would not only outshine those of the past in material terms, "they will elevate and ennoble man, lift him out of many of his present limitations, and make him the master where now he is the victim."[14] Engineers were convinced that history was on their side. As one of them put it, engineering "is the profession of the present, and will dominate the future."[15]

Associated with the sense of an exalted social role was a self-image of the engineer as a logical thinker. Engineers were assumed to have assimilated characteristics of science such as logic and impartiality. Thus, engineers concluded that they, as individuals, were peculiarly liberated by their profession from the usual human limitations, and therefore superior to other groups. One prominent electrical engineer maintained that the "vast majority" accepted "superficial, partial, and biased statements," concerning socio-economic questions, but that engineers, because they must live by "immutable laws" and must verify facts, think straight.[16] Another engineer spoke of the "position in society which belongs to us by right of education, achievement, and highly developed powers of logical deduction."[17]

The engineers' self-image as detached and impartial thinkers did not imply moral neutrality, however. Engineers assumed that they were morally as well as intellectually superior to other groups. Engineers were honest and altruistic lovers of truth. This concern for truth made them believers rather than skeptics. An ASME president, E. D. Meier, thought that "as we reverently discover and apply natural laws, we find new reasons and supports for . . . fundamental ethical conceptions."[18] Since the engineer lived by these "immutable laws," he learned honesty and morality in the course of his professional work. Another engineer, H. W. Buck, defended engineers and scientists from the charge of materialism. A world ruled by them would not be cold, materialistic, and atheistic. He thought that engineers possessed "exactly those qualities of mind and temperament best suited to combat materialism."[19]

Neither the self-image nor the social role that engineers ascribed to themselves bore close resemblance to reality. Engineers were not superhuman logicians, nor were they scientists. Their social thought was to reveal more than the usual human tolerance for contradiction. Although the scientific content of engineering practice was increasing, much that engineers did was not based on science. Indeed, in those areas of technology where science was coming to predominate, engineers were being displaced by natural scientists holding Ph.D. degrees, as in the budding industrial research laboratories. The engineers' claims, however, made sense in terms of professionalism. By stressing the esoteric quality of their work they laid the foundation for a claim for greater autonomy, since such activities could only be properly understood by initiates. The assertion that engineers were logical and impartial implied that if given greater freedom they would not misuse it. Professions must be self-regulating, and such internalized values are the most effective means for preventing the abuse of power. Inaccurate as literal descriptions, the engineers' self-portrayals expressed deeply felt aspirations for freedom and responsibility.

Accompanying the increasing self-consciousness of engineers was a mounting dissatisfaction with their status. This discontent was not new; a student of early American technological thought has reported an "almost morbid concern over the social status of the technologists" as early as the Jacksonian period.[20] But the new ideology invited unrest. Since engineers had built civilization, they ought to have a major share in its control, as well as greater recognition and honors. Engineers complained that their place was taken by lawyers, businessmen, and politicians, while the engineer was relegated to a subordinate position. The engineer was "a servant where he should be a master."[21] Especially should engineers control engineering work; the engineer should "possess his own."[22] In professional terms, engineers resented their lack of autonomy and the fact that engineering work was not controlled by colleagues.

The philosophy of professionalism carried engineers' ambitions beyond technology to politics and policy making generally. A mechanical engineer, Henry Hess, thought that in public affairs the engineer should have "a place more in accord with the importance of engineering."[23] Implicitly such men assumed that the hierarchy of professional excellence should apply to society at large and that position and power ought to be based on knowledge and skill. Pride in professional knowledge was fundamental to such thinking. An ASCE president based his hope for an increase in the status of the engineer on the fact that "the secrets of power" were in his keeping.[24]

One remedy suggested by engineers for their low status was greater professional loyalty. A president of the AIEE, C. O. Mailloux, suggested that the position of a group depended less on its merits than on its internal cohesion in struggling with other classes. He therefore urged engineers to develop more "guild spirit," which he defined as "that force which makes for the increase of prestige, influence, and power of the guild."[25] An ASCE president, George H. Benzenberg, maintained that the individual owed loyalty to his profession because individual successes were made possible by the collective labors of those colleagues who had patiently accumulated and disseminated the knowledge on which all engineering practice rested.[26] George S. Morison argued that the individual must subordinate himself to the group if the profession was to accomplish its mission.[27]

The philosopher's stone that was to transmute professional loyalty into higher status was social responsibility. Not by direct "lobbying" or selfish behavior, but through disinterested public service would the engineer enhance his position in society. Thus, C. O. Mailloux noted that the engineer's desire for more power and status was "in a sense" selfish. But he maintained that "fortunately, the real motive . . . is . . . quite altruistic, for the benefits which will result . . . for the engineering profession will be trifling in comparison with the benefits to the community, to the state, and to humanity in general."[28] Social responsibility was the master key to professionalism. The exercise of a sense of social responsibility implied independence. Conversely, an assertion of professional responsibility was a tactically effective means of advancing autonomy. It made such behavior a moral obligation for the engineer, while offering minimal affront to employers. Successful leadership would lead to the control of engineering work by engineers. Social responsibility, then, was a means whereby the group might gain power, independence, and social recognition.

But social responsibility was more than just a rationalization of selfish interest. It was an assertion of the value of the free individual, and it entailed a moral concern for the well-being of society. These were genuine and deeply felt commitments for some, if not all, engineers. Although expressed within a particular ideological framework, they were independent of that context. Some engineers found their ultimate justification for these beliefs in Christianity or the Declaration of Independence. Such effectiveness as the engineers did achieve rested on this sort of genuine moral dedication to values that transcended professionalism. That their success was less than their expectations was partially a consequence of mixing the values of responsibility and freedom with the baser alloys of hypocrisy, snobbishness, and self-interest.

The engineers' sense of stewardship over technology made them sensitive to public criticism of the misuse of their handiwork. At the turn of the century the public mood was confident. Engineers found their social responsibility a source of self-congratulation; they harbored few doubts about the beneficial results of their work. But the exposures of the muckraking journalists and the protests of progressive reformers shattered engineers' comfortable belief that technological progress automatically led to human benefit. Engineers became increasingly concerned as evidence of the misuse of technology mounted. In 1907, George H. Benzenberg urged fellow engineers not to attach themselves to "individuals or corporations engaged in monopolizing natural or artificial advantages for selfish purposes."[29] Some found in the abuses of capital a source of the engineer's low status; one professional spokesman complained that the engineer tended to become "the tool of those whose aim is to control men and to profit by their knowledge."[30] Philip N. Moore, a mining engineer, complained that many engineers were employed by corporations acting contrary to the public interest which "think they must hold their staffs to strict neutrality on all public questions."[31] Engineers urged their colleagues to assume a position of leadership and end the abuses of technology. An ASCE president wanted the engineer to "strive to become an engineer of men, pointing out lines of activity, based upon scientific principles, which permit of no discriminations or unfair advantages to favored interests."[32]

At the same time that engineers were beginning to chafe at the limits of their traditional technological role, the conservation movement was providing a possible alternative. Conservation was a movement in which scientists and engineers were playing a major role in guiding public policy. It was an example of central, scientific planning by technicians. Engineers engaged in resource planning were not limited to the perspectives of a single plant; they could take all the relevant variables into account. Thus, conservationists might hope to control the social consequences of technological change in a way not possible in ordinary engineering practice.

The Conservation Congress held at the White House in May of 1908 helped to crystallize the thinking of engineers concerning their profession's future role in politics. In a series of meetings before the main conference, the heads of the four major engineering societies drew up a series of resolutions that were later adopted by the conference. The most important of these was one calling for a department of public works to take charge of all of the federal government's engineering activities. Such an agency would apply to all government engineering the central direction characteristic of conservation. It was hoped that such a department would substitute rational calculation by experts for the usual political log-rolling. This proposal involved the creation of a cabinet post for an engineer, thus according the profession national recognition. No doubt a further motive was that of clipping the wings of the army engineers whose role in civilian engineering had long been resented. The conference and its attendant publicity convinced many engineers that if they organized and acted, their voices would be heard.[33]

The conservation movement was itself a part of the larger current of progressive reform. The attitude of most engineers toward progressivism was ambivalent. Engineers feared that society might be headed for disaster; but the worse conditions became, the more hopeful engineers were that they might save the situation by assuming social leadership. C. O. Mailloux regretted that the engineer's work had reacted on civilization so as to make necessary "profound alterations if not entire remodeling" of society. But by taking the lead in rebuilding, the engineer might greatly enhance his status.[34] Similarly, an ASCE president observed that society had "thrown away the old gods" without adopting new ones; and he, too, thought the engineer admirably fitted for the task of social reconstruction.[35] The First World War served both to intensify engineers' anxieties about the drift of events and to arouse their hopes that they might assume leadership in the postwar period. "The old order has failed and can never wholly return," proclaimed an electrical engineer in 1919. He urged engineers to guide society rather than "leave the steering of the ship to those we consider incompetent."[36]

At many points engineers' thinking paralleled that of contemporary progressive reformers. Perhaps surprisingly, engineers shared the progressives' nostalgia for a simpler, individualistic social order such as had once prevailed in rural and frontier settings. Like the progressives, engineers saw themselves as a middle group between capital and labor. Both were horrified at the possibility of class warfare. Above all, they shared a common faith, what George Mowry has called the "firm belief that to a considerable degree man could make and remake his own world."[37] Indeed both progressives and engineers drew much of their inspiration for this idea from the same sources: the past triumphs of technology, and the future potentialities of science. The engineers' version of this faith, however, was more restricted than that of most progressives. The hopes of engineers were focused on technological knowledge possessed by a body of initiates rather than general intelligence widely diffused through the population. These two views overlapped in the conservation movement, which could be seen variously as an effort by the people to control resource policy or as an example of scientific planning by experts.

The rhetoric employed by engineers was often strikingly similar to that of reformers. It misled even so sagacious an observer as Thorstein Veblen. But the similarities were deceptive. Despite a basic philosophical kinship, the differences were even more profound. Leaders of the engineering profession were acutely conscious of these differences. They regarded their proposals as substitutes for progressive reforms, not as supplements to them. Running through the statements of engineers on social questions was a hostility toward reformers which found expression in the occasional use of epithets such as "agitator," "utopist," "demagogue," and "dilettante."[38] Engineers tended to belittle the leadership of politicians, lawyers, and financiers, and to stress their own profession's qualifications for solving social problems. Their complaints about the drift of events often carried strongly conservative overtones. John A. Bensel in 1910 complained:

> Private ownership as it formerly existed is no longer recognized; individual action in almost any large field is today hampered and curtailed in a manner undreamed of twenty years ago. In fact, our whole scheme of government seems to be passing from the representative form on which it was founded, to some new form as yet undetermined.[39]

But such conservative tendencies were often masked by calls for seemingly drastic social reconstruction. Bensel did not preach a return to the old order; rather he saw a new one emerging. If engineers would become "cohesive," he thought that they might become the architects of this new order. He concluded that "the best doctors for our troubles are not necessarily those whose sympathies are most audibly expressed."[40]

The most fundamental difference between engineers and progressive reformers was their attitude toward democracy. Engineers did not share the progressives' faith in democracy. The driving force moulding engineers into a cohesive unit and pushing them into politics was an essentially elitist concept, professionalism. Professionalism involved concepts of social hierarchy and the inequality of man, whether in the membership grades of technical societies or in society generally. Engineers insisted that one man's opinion was not as good as another's. Conversely, spokesmen for the profession warned of the alarming "present disregard for knowledge and authority among the masses" and of the dangers stemming from the "less intelligent classes."[41] As a reform philosophy, professionalism stressed the creative role of the technical expert. Alexander C. Humphreys, president of the ASME in 1913, urged that reform should not depend on "hysterical suggestion" but on sane and impartial investigation, which engineers were best fitted to provide.[42] A later president of the ASME, Mortimer E. Cooley, thought that "the world today needs a Moses to lead it out of the Wilderness," and he was convinced that such a man would be an engineer.[43] Engineers showed no sympathy with such progressive staples as the initiative, referendum, and direct primary. It was to the elite of engineering professionals rather than to the wisdom of the masses that the engineers looked for social salvation.

Professional practice provided engineers with a pattern for the reconstruction of society. They rejected marginal adjustments between blocks and showed little sympathy with compromise. Instead, they sought scientific solutions to social problems. They assumed that there were laws governing human affairs that were fundamentally similar to those employed by engineers in their ordinary work. Some engineers seem to have imagined these laws were already known and identified them more or less vaguely with the "laws" of classical economics or social Darwinism. Other engineers assumed that such laws were as yet undiscovered. Comfort A. Adams asked:

> Are there no laws in this other realm of human relations which are just as inexorable as the physical laws with which we are so familiar? Is there no law of compensation which is the counterpart of our law of conservation of energy?[44]

Two assumptions reinforced engineers' belief in the possibility of "scientific" solutions to social problems. One was the notion that the laws of nature governed man and society. Thus, engineering could be assumed to include politics and economics as part of its subject matter. Overlapping this idea was a second: that the important thing was the nature of the method and the qualities of mind employed rather than the specific subject matter. Gano Dunn praised engineers' "way of thinking," which "enables us successfully to think of any kind of thing."[45] This line of reasoning led another electrical engineer, H. W. Buck, to assert that "it matters not whether the problems before him are political, sociological, industrial or technical, I believe that the engineering type of mind . . . is best fitted to undertake them."[46]

Although engineers were unclear as to the precise nature of the scientific laws governing human society, they were able to draw several clear inferences from their assumed existence. Scientific solutions to social problems would be permanent; once science solves a problem it need not be solved again. Thus, one engineer criticized existing methods for setting railroad rates as not in accordance with "the immutability of natural law."[47] Since engineering was the profession that applied scientific laws to practical problems, scientific solutions to social problems meant putting engineers in positions of leadership. Engineers assumed that such laws would be both material and moral; the engineering of society would lead to justice. An ASME president, E. D. Meier, urged the necessity of reform, since the abuses of capital had driven labor into "protest, dangerously near rebellion." He thought that "the remedy lies in placing engineers in all responsible positions in these great industries." The engineer, guided by scientific law, would find the "sane middle ground between grasping individualism and Utopian socialism." The result, Meier was convinced, would be that "the golden rule will be put in practice through the slide rule of the engineer."[48]

The contrasts between progressive reformers and engineers stemmed in part from differing social perspectives. Both groups were obsessed with status, but for different reasons. The progressive leadership was drawn from the old, independent middle class. They were small businessmen and independent professionals. They feared that the growth of large corporations, political machines, and big labor threatened to displace them from their position in American life. Thus, progressive reforms carried an anti-organizational bias. In contrast the engineers were exemplars of the new middle class of corporation employees. They had no fears of being displaced; the further growth of big business only opened up future opportunities for their profession. Rather theirs was a revolution of rising expectations. They wanted to better their position within the bureaucratic world of the corporation. They wanted recognition as professionals. They wanted a higher degree of individual autonomy. They wanted colleagues at the apex of the corporate pyramid. But they could not destroy big business without abolishing their own positions. They were fundamentally pro-organization in outlook.

Engineers differed from progressives in seeing no danger in the power and prerogatives of business; since they expected to inherit the corporate world, they did not want their patrimony diminished. Followers of Theodore Roosevelt's New Nationalism might agree with engineers that big business was part of a broad movement toward cooperation and efficiency. But where progressives favored regulation by government, engineers looked for reform coming from within the business community and through the agency of the engineer. Engineers accepted without question the structure, power, and basic ideological principles of business. What was needed was better management and more efficiency. These did not require government control of business, though such objectives did not preclude a milder form of cooperation between the two. Just how the engineer was to achieve these goals was seldom made clear. Few engineers presented anything that could be called a coherent program. But an obvious first step was to unite the fragmented engineering profession and direct it toward social problems.

In contrast to the engineers' sympathy with big business was their hostility toward organized labor. Some engineers were willing to grant a grudging and contingent acceptance of trade unions. Such organizations might have been necessary to protect the worker from grasping of incompetent management. But once engineers were in control and justice established, the unions would have no further function, and engineers looked forward to their ultimate disappearance. One leader of the ASME, George Melville, maintained that one of the social responsibilities of the engineer would be to reestablish "absolute freedom of contract."[49] Engineers aimed at the total elimination of class conflict under the aegis of science. The labor problem therefore was one for the engineer to solve. Calvert Townley, an electrical engineer, urged his colleagues to "dispel the bogy of class control. Brains always have ruled the world and brains always will. Show that we are dealing with a perfectly normal problem which must be solved in conformity to well-known natural laws."[50]

The engineers functioned politically in many ways as a part of a conservative reaction against progressivism. Their social indignation was not aroused by the injustices suffered by workers, farmers, and other disadvantaged groups. Rather it was the success of these underdogs in gaining redress that alarmed engineers, especially when labor unions and reformers seemed to threaten the existing structure of power and traditional social values. Serious concern among engineers arose in the period 1908 to 1912, paralleling the leftward shift of reform in the Insurgency movement. The Wilson administration aroused further anxiety and demands for action. Engineers did not effectively mobilize and act until after World War I, and their period of greatest activity, from 1919 to 1924, coincided with the nation's return to "normalcy." But engineers were not negativist followers of Harding and Coolidge; they saw themselves as reformers with constructive alternatives to existing reform programs. They wanted greater national efficiency, scientific planning, and more or less drastic administrative reforms of business and government. Their curious blend of conservatism and reform was to find national expression in the philosophy of the engineer-politician, Herbert Hoover.

It is perhaps surprising that there was virtually no effective opposition to the development of the philosophy of professionalism within engineering societies prior to 1920. There were a few dissenters, mostly representatives of an older generation. Rossiter W. Raymond, the autocratic secretary of the AIME, would have nothing to do with this new ideology. But he was of the same generation as Thurston; he had, in fact, founded the AIME before the ASME. And like Thurston, Raymond thought that the engineer was a businessman. Raymond was almost the only vocal opponent among the elite of engineering-society leaders. There were probably others, but they chose to remain silent. That is not to say that the engineers had achieved an undivided loyalty to their profession or a unanimous consensus on social issues. Quite the contrary. But those who were spokesmen for business chose to operate from within the ideological framework of professionalism.

Business-minded engineers readily embraced the philosophy of professionalism because it contained a fundamental inconsistency which enabled them to employ it for their own purposes. A complete doctrine of professionalism might be expected to insist on lifetime commitment on the part of the practitioner. This would follow logically from the exalted notion of the engineer's qualities, the importance of his social role, or the necessity of loyalty to the profession. In fact, a few engineers did draw this inference. Victor G. Hills, a mining engineer, argued that the engineer should view his profession as a lifetime commitment "and not as only a stepping stone in finance." He compared the commercial engineer to the lessee who exploits a mine solely for maximum immediate profit, neglecting safety precautions, leaving rubbish scattered about, and finally abandoning it to cave in. In contrast, the "thoroughbred professional man" was like the owner of a mine who does his work carefully, making provision for the future and for others.[51]

Hills's plea was almost unique; most professional leaders accepted and glorified the fact of the engineer's social mobility. Considering how such mobility was built into the very structure of their profession, engineers had very little choice in the matter; they simply made a virtue of necessity. Professional spokesmen linked the engineer's rising in the corporate hierarchy with his mission to transform civilization, establish justice, and end class conflict. Advocates of the ideology of professionalism maintained that engineers in such positions would still be practicing their profession. "It should not be considered unprofessional for an engineer to be a capitalist," maintained one ASCE president.[52] Such thinking made it all but impossible to distinguish between engineering and business management.

Spokesmen for professionalism assumed that the engineer in management would continue to be a loyal member of his profession and devoted to its values. They failed to foresee the possibility that professional loyalty might be used to undermine professional values. Where older engineers, like Thurston and Raymond, had argued that the engineer was a businessman, the newer generation of management-oriented engineers could maintain that the businessman was an engineer. Dugald C. Jackson, an educator affiliated by a lucrative consulting practice to the electrical utilities, took issue with a definition of the engineer presented to the AIEE in 1913. Any proper definition of the engineer, Jackson insisted, must include the executive officers of great corporations doing engineering work.[53] Since virtually all corporations were involved in engineering to some degree, this line of argument led to the total identification of the businessman and the engineer. "The engineer is a businessman," remarked George F. Swain, "for engineering is business and business is engineering."[54] Thus businessmen-engineers accepted the outer husk of professionalism while denying its inner essence, the sense of constituting a separate and independent group.

Business-oriented engineers like Swain and Jackson were even more insistent than other engineers, if possible, on the central role of the engineer in progress and his unique personal qualities. They could be so because the profession's acceptance of social mobility allowed them to use such doctrines to attach the engineer's loyalty to his employer. If engineers wanted to rise in the corporate hierarchy, they had to meet the demands of those who were in a position to confer or withhold promotion. Foremost among these requirements was loyalty. Thus, one president of the ASCE, John F. Wallace, stressed the need for the engineer to display "personal loyalty, not only to the enterprise in which he is engaged, but also to his immediate superior."[55] Such loyalty led to a more general commitment to the business system in general. As one civil engineer noted, "the successful engineer is the one who grasps the principle that engineering is business."[56] This loyalty was an obligation to the engineering profession, not just to the employer. The codes of ethics adopted by the AIEE in 1912 and the ASME in 1914 enjoined that "the engineer should consider the protection of a client's or employer's interests his first professional obligation."[57] In a real sense social mobility was subversive to professional loyalty. As an instrument of professional imperialism, it proved to be a double-edged weapon.

The intellectual counterpart of social mobility was success philosophy, a doctrine that entailed not only loyalty to employers but commitment to an ideology fundamentally at variance with professionalism. Success philosophy was very old; it had its origins in Puritanism, was secularized and reinforced by Franklin's philosophy of Poor Richard, and was popularized in the writings of Horatio Alger.[58] It asserted that the key to success lay in moral virtues such as hard work, self-denial, loyalty, enterprise, and initiative. But this contradicted the professional view of success, which maintained that sound engineering work rested on initiation in the esoteric knowledge painfully accumulated and preserved by the profession. Success philosophy stressed virtue, professionalism emphasized knowledge. The tension between the two was between "ought" and "is."

Success philosophy led to a view of society radically different from that implied in the engineers' ideology. Since success was a result of virtue, it followed that the successful were the virtuous; those who failed were morally deficient. Thus, society was viewed as a morality play in which the righteous were raised up and the unrighteous cast down. The scientific professionalism of the engineer, in contrast, stressed the importance of material factors in shaping society. Its adherents asserted the desirability of manipulating the environment in accordance with scientific understanding. This conflicted with success philosophy which implied that what is, is right, and that any attempt to change the status quo by collective action would be immoral. Professionalism attributed progress to the increase of technological knowledge; success philosophy related material advances to the practice of virtue. Both doctrines were, at base, elitist; but success philosophy looked to an aristocracy of wealth and virtue, while professionalism implied a technocracy of knowledge and skill.

The most important conflict between success philosophy and professionalism came over the issue of social responsibility. From the moralist viewpoint posited by success philosophy, the ills of society must be due to an absence of virtue. The remedy lay in moral reform. Thus, President Edwin S. Carman of the ASME feared that irresponsibility and greed threatened to wreck society, and he urged engineers to remedy this situation by discovering and enforcing means for securing "complete conformity" to moral law.[59] The moral virtues stressed by engineers were usually those associated with capitalism. This led to a theory of social responsibility, that of defending business. Alexander C. Humphreys maintained that it was the engineer's duty to defend business, particularly by educating the public on the matter of excessive governmental controls.[60] Such a responsibility, since it did not rest on esoteric knowledge unique to the engineer, might best be fulfilled by joint action with the business community. In 1915, E. W. Rice, president of both the AIEE and of General Electric, noted that "it looks as if our businessmen now propose to make a business of seeing to it that they are properly represented in the business of government," and he urged engineers to join "this great movement."[61]

The ambiguous use of the term "law" by engineers was another means of converting professionalism into a doctrine of business loyalty. Like social mobility, it was stressed by professional leaders as a means whereby engineers might gain control of engineering work. The scientist-engineer emphasized the priority of physical laws; but without the ambiguous usage of "law" he would not have been able to assume that human relations, economics, and politics were part of the subject matter comprised within engineering, or that "engineers" engaged in management and administration were colleagues. Businessmen-engineers readily accepted this usage of "law" and its consequences. But they could argue that it was the engineer's duty to uphold such "laws" as those of classical economics and social Darwinism. D. C. Jackson maintained that the modern corporation was the product of an evolutionary law discovered by Spencer and, therefore, that it was the engineer's duty to defend such corporations from government regulation.[62] A further consequence of the dilution of the idea of "law of nature" was to weaken the profession's claim to autonomy. The classic argument has been that outsiders could not judge the professional's work, since it involved esoteric knowledge known only to initiates. But if businessmen were members of the same group and utilized essentially the same knowledge as the engineer, it would not be tenable to argue that businessmen could not understand engineering nor control engineers. Thus, E. W. Rice spoke of "the simple, common-sense methods used by engineers and successful businessmen in the ordinary course of business."[63]

Engineers were only dimly aware that there was any contradiction in their social thought. They did not accept one set of ideas to the exclusion of the other, but tried to combine them. As a result, the thinking of engineers on social questions has oscillated between the poles of "ought" and "is." To some degree engineers achieved consistency by applying differing viewpoints to different classes in the community. They have thought of themselves and businessmen as free moral agents, but, as William E. Wickenden observed, engineers have tended to think of workers and consumers in materialistic terms.[64] George F. Swain in 1913 maintained that the poor should not envy the rich, but in accordance with the Christian ideal he wanted "kindliness and a spirit of brotherhood between all men." But later in the same address, when considering the proper attitude that the rich ought to take toward the poor, he questioned the desirability of philanthropy, since in evolutionary terms it enabled the unfit to survive. He believed that "if present-day humanitarianism leads to race degeneration, it cannot be the way of progress intended by an all-wise Creator," and quoting Paul, he urged that those who did not work should starve.[65]

The notion, derived from evolution, that values are relative to changing social conditions illustrates the way engineers have tended to employ concepts. When faced with ideas that they did not accept, they have applied a relativistic analysis. Thus, engineers have frequently favored cooperation rather than competition in business, and large enterprises over small ones. Such ideas as atomistic competition may then be treated as relative to a particular time and place. But when their own values are challenged, they are likely, with Arthur P. Davis, president of the ASCE in 1921, to stand on absolute moral principles, since "no citadel is invincible except the citadel of righteousness."[66]

Where moralistic and scientific systems of thought led engineers to different answers to the same questions, a verbal accommodation might be achieved by simply combining the two. Thus, some engineers argued that progress depended on both technological knowledge and moral virtue. George F. Swain thought that civilization was mainly due to the work of the engineer and that continued progress would require initiative, hard work, private property and like values.[67] Moral reform and scientific planning could be related as ends and means. Because engineers used moral and scientific laws interchangeably, the difference between planning and restoring morality was obscured. Where a clear-cut difficulty arose was in the fact that planning necessitated collective action. This could be resolved by emphasizing that the engineer's own collective actions would have as their end the restoration of traditional moral values. George H. Pegram, president of the ASCE in 1918, thought that the engineer's work was endangered by a "wave of collectivism." To combat this threat he wanted engineers to profit from the example of the doctors and lawyers and form an "alliance of all engineering interests." The end of such collective action, however, would be the defense of individualism; Pegram thought that "if individualism is not protected, we must drift toward the jungle."[68]

This purely verbal reconciliation between the competing ideologies of engineering and business, however, proved effective only so long as engineers confined themselves to words. When, after 1908, they moved from words to deeds, the underlying tensions in their thought became manifest in differing proposals for action. To the businessman-engineer the way to progress was through the operation of beneficent laws of an essentially moral character, operating through the private-enterprise system. To assure progress the engineer, side-by-side with the businessman, should fight to prevent any tampering with the operation of these laws by unwise legislation. In short, the engineer should be a moralist. His responsibility would be that of defending individualism, initiative, and other virtues against the collectivist tide. In contrast, a scientific orientation placed the engineer in the role of a social planner. Since society was governed by physical laws like those of science, the solution to social problems would be found by manipulating the environment in accordance with these principles and by means of the engineering method. Since the solutions were to be found in science, this line of thinking led engineers to the idea of independent action by their own profession, rather than to joint action with the business community. No mere verbal manipulation could bring these contrary conceptions of social responsibility into complete harmony.

At base the engineer's problem was not logic but bureaucracy. The independent role for the profession, implicit in professionalism, represented a rebellion against bureaucracy. Even if the ends sought were the defense of business, this would still be true; engineers would be acting outside the bureaucratic chain of command. Professionalism was a means of limiting bureaucratic control by increasing the group's autonomy and the degree to which it controlled its own work. Any effort to implement these goals would meet with the opposition of the defenders of the prerogatives of business, however much agreement might be sought on the verbal level. Indeed, once engineers challenged the entrenched power of business, they produced a counteraction on the part of business to control the engineering profession. Such was to be the consequence of professionalism after 1920.

NOTES

1. "Symposium on 'The Status of the Engineer,' " *Trans AIEE*, XXXIV, pt. 1 (January–June, 1915), 324.

2. Robert H. Thurston, "President's Inaugural Address," *Trans ASME*, I (1880), 15.

3. Charles H. Loring, "The Steam Engine in Modern Civilization," *Trans ASME*, XIV (1892–1893), 52–58.

4. "How Can the Present Status of the Engineering Profession be Improved?" *ibid.*, 497–501. For a similar effort, see Benjamin M. Harrod, "Address at the Annual Convention," *Trans ASCE*, XXXVII (January–June, 1897), 540–542.

5. Robert H. Thurston, "The Mechanical Engineer—His Work and His Policy," *Trans ASME*, IV (1882–1883), 96–98, and his "President's Inaugural Address," *Trans ASME*, I (1880), 3–5, 8.

6. Charles H. Loring, "The Steam Engine in Modern Civilization," *Trans ASME*, XIV (1892–1893), 52–58.

7. A. L. Holley, "The Field of Mechanical Engineering," *Trans ASME*, I (1880), 1–6.

8. George S. Morison, "Address at the Annual Convention," *Trans ASCE*, XXXIII (June, 1895), 467.

9. *Ibid.*, 483.

10. *Ibid.*, 474.

11. Charles P. Steinmetz, "Presidential Address," *Trans AIEE*, XIX (January–July, 1902), 1145.

12. Gano Dunn, "The Relation of Electrical Engineering to other Professions," *Trans AIEE*, XXXI, pt. 1 (January–June, 1912), 1031.

13. Charles Hermany, "Address at the Thirty-Sixth Annual Convention," *Trans ASCE*, LIII (December, 1904), 464.

14. George H. Babcock, "The Engineer: His Commission and His Achievements," *Trans ASME*, IX (1887–1888), 37.

15. E. D. Meier, "The Engineer and the Future," *Trans ASME*, XXXIII (1911), 495.

16. Comfort A. Adams, "Facts Versus Propaganda," *Electrical Engineering*, LX (August, 1941), 371–373. Although this statement is after 1920, Adams was president of the AIEE in 1919. He expressed similar sentiments then, if less graphically (see Comfort A. Adams, "Cooperation," *Trans AIEE*, XXXVIII, pt. 1 [January–June, 1919], 790, 793).

17. Henry Gordon Stott, "The Evolution of Engineering," *Trans AIEE*, XXVII, pt. 1 (January–June, 1908), 459.

18. E. D. Meier, "The Engineer and the Future," *Trans ASME*, XXXIII (1911), 495.

19. H. W. Buck, "The Engineer's Destiny," *Trans AIEE*, XXXVI (January–June, 1917), 601.

20. Hugo Arthur Meier, "The Technological Concept in American Social History, 1750–1860" (Ph.D. dissertation, University of Wisconsin, 1950), 498.

21. Onward Bates, "Address at the Forty-First Annual Convention," *Trans ASCE*, LXIV (September, 1909), 573.

22. Hunter McDonald, "Address at the Annual Convention," *Trans ASCE*, LXXVII (December, 1914), 1755.

23. Henry Hess, "Relation of Engineering Societies to Legislation and Public Affairs," *Journal ASME*, XXXVIII (August, 1916), 641.

24. Robert Moore, "The Engineer of the Twentieth Century," *Trans ASCE*, XLVIII (August, 1902), 233.

25. C. O. Mailloux, "The Evolution of the Institute and of Its Members," *Trans AIEE*, XXXIII, pt. 1 (January–June, 1914), 828.

26. George H. Benzenberg, "The Engineer as a Professional Man," *Trans ASCE*, LVIII (June, 1907), 520.

27. George S. Morison, "Address at the Annual Convention," *Trans ASCE*, XXXIII (June, 1895), 476, 483.

28. C. O. Mailloux, "The Evolution of the Institute and of Its Members," *Trans AIEE*, XXXIII, pt. 1 (January–June, 1914), 831.

29. George H. Benzenberg, "The Engineer as a Professional Man," *Trans ASCE*, LVIII (June, 1907), 521.

30. Onward Bates, "Address at the Forty-First Annual Convention," *Trans ASCE*, LXIV (September, 1909), 573.

31. Philip N. Moore, "The Civic Responsibility of the Engineer," *Mechanical Engineering*, XLI (May, 1919), 448.

32. Charles Macdonald, "Address at the Fortieth Annual Convention," *Trans ASCE*, LXI (December, 1908), 546.

33. *Ibid.*, 548–551; Henry G. Stott, "The Evolution of Engineering," *Trans AIEE*, XXVII, pt. 1 (January–June, 1908), 462–463.

34. C. O. Mailloux, "The Evolution of the Institute and of Its Members," *Trans AIEE*, XXXIII, pt. 1 (January–June, 1914), 831.

35. John A. Bensel, "Address at the Forty-Second Annual Convention," *Trans ASCE*, LXX (December, 1910), 468.

36. Comfort A. Adams, "Cooperation," *Trans AIEE*, XXXVIII, pt. 1 (January–June, 1919), 783.

37. George E. Mowry, *The Era of Theodore Roosevelt and the Birth of Modern America* (New York, 1962), 37.

38. C. O. Mailloux, "The Evolution of the Institute and of Its Members," *Trans AIEE*, XXXIII, pt. 1 (January–June, 1914), 831, 833, and Henry G. Stott, "The Evolution of Engineering," *Trans AIEE*, XXVII, pt. 1 (January–June, 1908), 463.

39. John A. Bensel, "Address at the Forty-Second Annual Convention," *Trans ASCE*, LXX (December, 1910), 468.

40. *Ibid.*, 468–469.

41. C. O. Mailloux, "The Evolution of the Institute and of its Members," *Trans AIEE*, XXXIII, pt. 1 (January–June, 1914), 834; George Fillmore Swain, "Some Tendencies and Problems of the Present Day and the Relation of the Engineer Thereto," *Trans ASCE*, LXXVI (December, 1913), 1115.

42. Alexander C. Humphreys, "The Present Opportunities and Consequent Responsibilities of the Engineer," *Trans ASME*, XXXIV (1912), 630.

43. Mortimer E. Cooley, "A Survey of the Society's Organization," *Trans ASME*, XLI (1919), 590.

44. Comfort A. Adams, "Cooperation," *Trans AIEE*, XXXVIII, pt. 1 (January–June, 1919), 792.

45. Gano Dunn, "The Relation of Electrical Engineering to Other Professions," *Trans AIEE*, XXXI, pt. 1 (January–June, 1912), 1034.

46. H. W. Buck, "The Engineer's Destiny," *Trans AIEE*, XXXVI (January–June, 1917), 600.

47. Charles Macdonald, "Address at the Fortieth Annual Convention," *Trans ASCE*, LXI (December, 1908), 546.

48. E. D. Meier, "The Engineer and the Future," *Trans ASME*, XXXIII (1911), 497–498.

49. George W. Melville, "The Engineer's Duty as a Citizen," *Trans ASME*, XXXII (1910), 530.

50. Calvert Townley, "The Engineer, Employer and Employee," *Journal AIEE*, XXXIX (March, 1920), 237.

51. Victor G. Hills, "Professional Ethics," *Trans AIME*, XLI (1910), 560–561.

52. Onward Bates, "Address at the Forty-First Annual Convention," *Trans ASCE*, LXIV (September, 1909), 574.

53. William McClellan, "A Suggestion for the Engineering Profession," *Trans AIEE*, XXXII, pt. 2 (May–December, 1913), 1281.

54. George Fillmore Swain, "Some Tendencies and Problems of the Present Day and the Relation of the Engineer Thereto," *Trans ASCE*, LXXVI (December, 1913), 1141.

55. John Findley Wallace, "Address at the Annual Convention," *Trans ASCE*, XLIII (June, 1900), 606.

56. Maurice G. Parsons, "The Philosophy of Engineering," *Trans ASCE*, LXXVII (December, 1914), 45.

57. "Report of Committee on Code of Ethics," *Trans ASME*, XXXVI (1914), 23, and "Code of Principles of Professional Conduct of the American Institute of Electrical Engineers," *Trans AIEE*, XXXI, pt. 2 (June–December, 1912), 2227. The ASME code omitted the word "professional."

58. Irvin G. Wyllie, *The Self-Made Man in America* (New Brunswick, N.J., 1954), and John G. Cawelti, *Apostles of the Self-Made Man* (Chicago and London, 1965), are recent studies of success philosophy.

59. Edwin S. Carman, "Presidential Address," *Trans ASME*, XLIV (1921), 471.

60. "Symposium on 'The Status of the Engineer,' " *Trans AIEE*, XXXIV, pt. 1 (January–June, 1915), 318–319. See also Alexander C. Humphreys, "The Present Opportunities and Consequent Responsibilities of the Engineer," *Trans ASME*, XXXIV (1912), 633.

61. "Symposium on 'The Status of the Engineer,' " *Trans AIEE*, XXXIV, pt. 1 (January–June, 1915), 308.

62. Dugald C. Jackson, "Electrical Engineers and the Public," *Trans AIEE*, XXX, pt. 2 (April–June, 1911), 1137–39.

63. "Address by President-Elect E. W. Rice Jr.," *Trans AIEE*, XXXVI (January–December, 1917), 605.

64. William E. Wickenden, "The Social Sciences and Engineering Education," *Mechanical Engineering*, LX (February, 1938), 148.

65. George F. Swain, "Some Tendencies and Problems of the Present Day and the Relation of the Engineer Thereto," *Trans ASCE*, LXXVI (December, 1913), 1116, 1139–40.

66. Arthur P. Davis, "Federated Engineering Societies of America," *Trans ASCE*, LXXXIV (1921), 143.

67. George F. Swain, "Some Tendencies and Problems of the Present Day and the Relation of the Engineer Thereto," *Trans ASCE*, LXXVI (December, 1913), 1114, 1115.

68. George H. Pegram, "Address at the Annual Meeting," *Trans ASCE*, LXXXII (1918), 161, 165.

4 ▪ THE POLITICS
OF STATUS

In the first two decades of the twentieth century, American engineering societies were the scene of a power struggle largely unknown to the general public and only dimly understood by most engineers. At issue was professionalism. Groups of reformers sought to orient engineering societies to their profession, to unite them, and to make them an active force in politics. That is, they wanted to put the ideology of engineering into practice. In the technical press and in discussions before engineering societies, the reformers were sometimes termed "progressives" or "insurgents," and their opponents were called "conservatives" or "standpatters." The implied parallel between the turmoil within the engineering profession and the contemporary progressive reform movement in state and national government was not entirely inaccurate. Although engineering reformers differed from national progressives on many issues, they were animated by similar motives. Both were reacting to the threat to their status posed by a corporate, bureaucratic society. [1]

Like the national movement, engineering progressivism underwent a shift to the left in the period from 1910 to 1916. The first wave of engineering reform was led by a comparatively small professional elite among electrical and mining engineers. Although electrical and mining engineers initiated important professional innovations, they were unable to carry them to completion. Fundamentally, they suffered from the defects inherent in the ideology of engineering. Strongly elitist, they failed to generate mass support either among rank-and-file engineers or the general public. Obsessively concerned with status, they were unable to go from ceremonial gestures to the formulation of a coherent program for professional welfare and political reform. When faced with concrete problems, they were unable to justify the sweeping claims that they were making for the engineering mind. A craving for deference alone did not suffice.

A second, and more successful, phase of engineering reform was initiated by civil and mechanical engineers, particularly after 1915. But, however much the later progressives differed from their forerunners among electrical and mining engineers, they shared the same core of professional and status aspirations. At base, they were part of a single social movement.

Electrical engineers were in the vanguard of professional reform from 1900 to 1912. The reasons for their early preeminence are clear: electrical engineering, of all technical specialties, was at once the most professional and the most threatened by bureaucratic control. The development of the electrical industry helps to explain this apparent contradiction. In its early development the electrical industry was highly competitive, rather chaotic, and in need of advanced technological innovations. Under these circumstances, a high degree of professional autonomy was possible. But by 1900, the needs of the industry were changing. Technological innovation was no longer as crucial. The earlier internal divisions within the industry had largely disappeared through merger and consolidation. More than any other technical field, electrical engineering after 1900 operated in a field controlled by a small number of large corporations. Two lines of policy were open to the AIEE's leaders. They could further buttress professionalism as a counterweight to bureaucracy, or they could seek closer alliance with industry. The reformers chose the former.[2]

The symbol of the AIEE's professional spirit was its high standard for membership. Prior to 1912, the institute had only two membership grades, member and associate; those who joined the institute had to become associates first and then apply for full membership. The original rules, adopted in 1884, provided that full members should be professional electrical engineers; but they did not specify the meaning of "professional engineer."[3] This left it to the board of directors to determine the actual criteria to be met. In the early 1890s, the board adopted the practice of transferring to full membership no more than five persons per meeting, or a maximum of fifty annually. This was done "in order that such membership become year by year of constantly increasing value."[4] As a result the de facto requirements tended to rise steadily.

In 1901, the AIEE codified its criteria for membership in a new constitution. Except for age, these requirements were fully equal to those of the ASCE. A full member had to be both qualified to design and in responsible charge of electrical engineering work. The new constitution also restricted the offices of president and vice-president to full members. In a sense the change was purely symbolic, since all members could vote; and except for holding the two highest offices, they had equal privileges.[5] The meaning of the new constitution lay in prestige politics; it gave a dominant position within the institute to a small professional elite. Despite a lively debate, the new constitution was adopted at the annual meeting in 1901 with only three dissenting votes.[6]

Another factor that reinforced professionalism early in the twentieth century was the institute's unique system of open and competitive elections. When the institute was small, nominations and elections had been conducted at the annual meeting in a democratic, if sometimes rather chaotic, manner. In 1892, with increasing size and geographical dispersion of members, the AIEE adopted a letter ballot and authorized the board of directors to prepare an official slate of candidates to be presented along with other names suggested by the membership.[7] This did not, however, lead to the usual closed system of nominations and elections. One reason it did not was that in 1893 the official nominee for the presidency was defeated by a group of members who employed "Tammany methods" to promote a rival candidate.[8] This event helped to establish a precedent: the board of directors adopted the practice of always selecting as the official candidates those who had received the largest number of nominating ballots. In effect, the primary became the deciding election. Since every member automatically received a nominating ballot, these elections were entirely open and each office was contested.[9]

The AIEE's electoral system was far from perfect. Nominations tended to be scattered over a large number of candidates. In 1908, for example, there were 77 nominations for president, 258 for vice-president, and 524 for director.[10] Thus, the person with the largest number of votes was not necessarily the majority choice. Since the board did not attempt to function as a nominating committee, the electoral results did not produce geographical balance. New York was disproportionately represented on the board of directors because the largest block of members was concentrated in that city. But for all its confusion, the system led to more membership control of the AIEE than existed in any other major engineering society. It had the effect of raising to high office young men who had distinguished themselves professionally. Thus, the electoral system furthered the trend toward a professional orientation of the institute. Conversely, this system served to prevent the domination of the AIEE by any given industry or clique.

An outstanding example of the young professionals elevated to high office by the AIEE was Charles F. Scott. Elected president in September, 1902, Scott was typical of the scientifically trained young men who dominated electrical engineering in the early twentieth century. After studying physics at Ohio State under T. C. Mendenhall, he did graduate work in mathematics and physical science at Johns Hopkins. Scott rose rapidly in his profession. He became "chief electrician" for the Westinghouse Company at age thirty-three and was only thirty-eight when elected president of the AIEE. A pioneer in the development of alternating-current power, Scott gained fame with colleagues by his discovery of a means of transforming between a three-phase alternating-current system and a two-phase system by the use of two ordinary transformers. Like his immediate predecessor as president of the AIEE, Charles P. Steinmetz, Scott was interested in developing professionalism. His enthusiastic advocacy of engineering unity may have been a factor in his election. [11]

Scott wanted to do something about the inadequate status of the engineer. His remedy was "cooperation." This, he thought, had been the great discovery of the nineteenth century: "the effectiveness of concentration, the efficiency of largeness." [12] The engineer had provided the fundamental innovations which had made cooperation possible in industrial and financial affairs. Yet, the engineer lacked the recognition he deserved. The reason for this was that engineers were disunited and did not know their own strength. Scott suggested that engineers apply to themselves the same principle of cooperation that had worked so well in industry. He wanted engineers to work together to advance their own profession. "The men who are mastering the powers of nature will yet rise in the strength of united effort to meet the increasing responsibilities of the coming years," he maintained. [13]

Scott hoped ultimately to unite all engineers in a "national engineering congress" in which each society would preserve its separate identity, but in which all engineers would cooperate in advancing their profession. As a practical first step toward this goal, Scott advocated the creation of a joint headquarters building in New York for the four founder societies. [14] This idea was far from new; it had been discussed intermittently since the 1880s. By 1902, this issue had taken on a new urgency for the AIEE. The year previously a prominent member, Schuyler S. Wheeler, had purchased a large library of the literature of electrical engineering and presented it to the institute. But the AIEE had no headquarters building and no place for the library. Andrew Carnegie agreed to provide additional funds for the upkeep of the library, but he balked at giving the AIEE a permanent headquarters. Scott was able to put the entire matter in a larger perspective. At a dinner honoring Carnegie for his support of the library, Scott set forth his vision of a common headquarters building as an instrument for promoting cooperation among American engineers. Carnegie was interested in this suggestion, and the next day he invited Scott and Calvin Rice, a member of the institute's building committee, to call on him and discuss the matter. As a consequence of this meeting, Carnegie offered to give the founder societies one million dollars to translate Scott's dream into reality.[15]

Carnegie's gift came at a propitious time. The sentiment for professional unity was growing. The AIEE had no permanent headquarters, and both the ASME and AIME had outgrown their existing homes. Only the ASCE, which had just acquired a new building worth a quarter million, was not in need of a new headquarters. Purely as a business proposition, the ASCE had everything to gain by accepting Carnegie's offer, since they could sell their existing building and convert the proceeds into endowment. But a number of members, led by ex-presidents John F. Wallace and J. J. R. Croes, opposed any collaboration with the other major engineering societies. They thought that the term "civil engineering" included every field and that the ASCE was the proper unity organization for all engineers. Secondly, they argued that the lower entrance requirements of some engineering societies made them commercial rather than professional organizations. When the matter was referred to a letter ballot, the proposal for a common headquarters was defeated by a vote of 1139 to 662.[16]

Despite the intransigence of the ASCE, the other three societies went ahead with the construction of the new building. In at least one respect the plan was less broad and generous than that proposed by Scott and supported by influential segments of the technical press. He had suggested that the smaller national engineering societies also be allowed to share space in the Engineering Societies Building.[17] This was not done. As Scott had foreseen, the necessity of joint planning for the new building made further cooperation necessary. To manage the purchase of land, the approval of plans, and to supervise construction, the three societies organized the Joint Conference Committee. Initially a temporary expedient for a single purpose, this committee became a regular means for joint action by the founder societies. Before the Engineering Societies Building was completed, Gano Dunn, the president of the AIEE, suggested that the committee be made permanent.[18] It acted sporadically between 1911 and 1917 to express opinion on matters of public policy. To manage the new building when finally completed, the founder societies, by this time joined by the ASCE, created another joint organization, the United Engineering Society. Although primarily a housekeeping agency, its constitution was broad enough to include other forms of cooperation.[19]

Scott's ultimate aim was to unite all engineering societies, not just the larger national ones. In order to bring local engineering societies into "one great system" with the founder societies, he suggested the creation of local sections, which would work out cooperative arrangements with other local organizations.[20] This idea, too, was an old one and had been discussed since the 1880s, not only as a means of cooperation but also as a step toward a more democratic organization for engineering societies. Scott's proposal was adopted by the board of directors, and some thirty sections were formed the first year. Scott was enthusiastic. "Think for a moment," he wrote, "of the possibilities in elevating the work and increasing the efficiency of engineering organizations by establishing close ties between local and national societies."[21]

Besides unity, another means for enhancing the status of the engineer was the adoption of a formal code of ethics. Here again the AIEE took the lead; it was the first of the founder societies to adopt such a measure. Schuyler S. Wheeler, president of the AIEE in 1906, called for action on this long-discussed question. Like Scott, Wheeler was chiefly interested in advancing the "importance, dignity, and strength" of his profession.[22] In proposing a code of ethics, Wheeler showed little concern for specific abuses. Rather, he noted that

> the public finds no such declaration of high standards in the engineering profession as in the older professions. It is therefore inclined to treat the engineer lightly in ethics, and to withhold from him that deference which, if rendered, would in itself encourage a higher professional standard.[23]

Wheeler's proposal was seconded by Charles P. Steinmetz, who thought that such a code was particularly necessary for electrical engineers since most of them were "more or less closely associated with large manufacturing or large operating companies."[24]

The AIEE's code of ethics, finally adopted in 1912, bore little relation to industrial realities. It was phrased as if the engineer were a consultant. While dealing with such issues as conflicts of interest and the ownership of engineering plans, it ignored problems concerning engineers employed in a bureaucratic setting.[25] Wheeler assumed that loyalty to the client was primary "and the one to which all others must give way if there is any conflict."[26] The code adopted made loyalty to the client or employer the engineer's first professional obligation. Wheeler thought that the engineer's duty to the public was "largely educational."[27] The code prohibited criticism of fellow engineers and forbade the discussion of engineering subjects in the public press. These provisions, in fact, made it more difficult for an individual to take the side of the public in matters involving engineering. In any case, the AIEE's code of ethics was largely ceremonial, and no serious efforts were made to enforce it. Its aim was to enhance the status of engineers, not to improve their morals.

Another area in which the AIEE took the lead was public policy. In 1905, the institute authorized a committee on forest preservation. The scope of this committee was gradually enlarged until, by 1911, it included the entire field of public policy. Paralleling this interest in political questions, the AIEE expanded its work in standardization. For example, the institute participated in drawing up a national electric code. In public policy a wide range of subjects received committee attention, including conservation legislation, reform of the patent system, the role of local sections in civic matters, the regulation of electric utilities, licensing laws for engineers, and many other questions of political importance. Although the institute's board of directors was reluctant to take a positive stand on these and other issues, the new concern of the AIEE with politics was apparent in the publication of a number of papers dealing with such policy matters as conservation and utilities regulation. In politics, as with unity and ethics, the AIEE pioneered a significant broadening of the scope of engineering societies along lines suggested by a developing ideology of engineering.[28]

The inability of elite electrical engineers to go beyond mere status seeking was nowhere more apparent than in the field of public policy. One of the most fundamental ideological claims of engineers was the notion that there was a unique engineering approach to social problems. But the proposals of electrical engineers represented no more than a weak echo of the policies being advocated by the electrical industry. The institute's standardization work was done in close collaboration with the major electrical trade associations and scarcely represented an independent exercise of the engineering mind.[29] The AIEE refused to collaborate with existing conservation agencies, and it adhered to a narrow defense of the specific interests of the electrical industry.[30] For example, Lewis B. Stillwell presented a detailed criticism of the Forest Service's policy of collecting fees for the use of water-power sites. His argument was that "any plan which imposes a tax on water power or fuel is at best of doubtful wisdom."[31] Papers on utilities regulation were presented by men affiliated with industry and offered no new ideas. Mortimer E. Cooley, for example, denied that any real problem existed, and he blamed the "self-appointed guardian of the people's interest" for most difficulties.[32]

The AIEE's public-policy program provided two options. On the one hand, elite electrical engineers offered no unifying ideas, no comprehensive programs, no clarion calls that might rally support. In fact, they offered little that was new at any level. Thus, the institute's activities aroused little or no interest among engineers or the general public. But on the other hand, despite their impeccable orthodoxy, the AIEE's activities posed a potential threat to business interests. The institute enjoyed a high degree of autonomy, and there was no guarantee that its position would continue to coincide with that of the electrical industry. A few prominent members held heterodox ideas. Charles P. Steinmetz was a socialist and an advocate of national planning.[33] Ralph D. Mershon, a consulting engineer and inventor, supported the federal government's policy of short-term licensing for water-power sites. Spokesmen for the utilities argued that the existing licensing system increased the cost of electric power to the consumer. Mershon attacked this contention, and he went on to defend the government's policy as in the public interest.[34] Mershon's election as president of the AIEE in September, 1912, could scarcely have been reassuring to utilities interests.

Steinmetz and Mershon, however, were exceptions. The men who controlled the institute were, for the most part, soundly conservative. So long as the power structure of the AIEE remained unchanged, the threat posed by the institute's political activities was a very remote one. But the growing power of the local sections raised the possibility of a major restructuring of the AIEE. Since the aristocracy of the profession tended to congregate in a few metropolitan centers, decentralization would tend to increase the importance of ordinary engineers at the expense of the elite. And while the institute prior to 1912 was not dominated by businessmen, they did have a voice in its affairs. Business interests benefited from the concentration of power in New York. A decentralized society would be harder for them to influence.

The local sections, as initially set up, had little independence and were only authorized to discuss papers presented at national meetings. President Scott, however, had granted the sections a great deal of autonomy in their own internal affairs.[35] Within a few years, they were demanding not only more freedom but also a larger say in institute government.[36] At first the board of directors was willing to make concessions. In 1910, the board authorized the sections to hold their own meetings. They were also allowed to work out cooperative arrangements with other local societies.[37] This carried far-reaching implications. In reform-dominated cities, such as Cleveland, local engineering societies were sometimes closely allied with progressive political factions, to the point of supporting such radical measures as municipal ownership of utilities.[38]

The movement to decentralize the AIEE came to a head in 1912. The board of directors acted to curb the political independence of the sections. It adopted a bylaw defining the sections' work in narrowly technical terms. It forbade any action or publication that might commit the institute or the section on matters of policy without prior approval of the board.[39] The sections countered by demanding that the special position of New York members be abolished. The St. Louis section drew up a chart showing that New York, with 19.4 percent of the membership, filled 52.5 percent of the institute's offices. The St. Louis members proposed that offices be distributed on a geographical basis; and to ensure attendance, they suggested that the institute pay the expenses of officers resident outside New York. They wanted New York meetings reduced to the same position as meetings held in any other city.[40] At an AIEE meeting in June, 1912, the delegates of the local sections discussed these proposals. They adopted a resolution requesting that the board of directors appoint a special committee to draw up rules allowing the institute's sections "to take a larger part in the conduct of its affairs."[41]

The AIEE's growing interest in public policy and the incipient revolt of the local sections, while innocuous enough in themselves, could be very dangerous if they led to an alliance between electrical engineers and urban reformers. The real threat to business interests came from the popular movement for government ownership or stringent control of utilities. Public service corporations, because of their relation to municipal corruption, had become principal targets of progressive reformers. In the period 1907 to 1911, both the National Electric Light Association and the Bell telephone interests came to accept state regulation as a lesser evil than government ownership.[42] Beginning in 1907, with Wisconsin, state after state set up regulatory commissions where engineering knowledge often played a crucial role. Thus, after 1907, it became a matter of great importance to the utilities that the engineering profession be allied with them rather than with progressive reformers.[43]

Engineers affiliated with the utilities began to call for solidarity between the AIEE and the utilities industry. In 1908, the institute elected L. S. Ferguson as its president. He was a vice-president of Commonwealth Edison of Chicago and an associate of Samuel Insull. According to *Engineering News*, Ferguson wanted to "interest those men who direct large industrial work and who although not directly engaged on the technical side, could advise and assist very materially in the plans of the Institute."[44] He also hoped to see closer cooperation between the AIEE and the electrical trade associations, such as the NELA. In 1911, President Dugald C. Jackson, a consultant allied with the utilities, argued that engineers had a moral obligation to defend utilities threatened by unfair public regulation.[45]

The AIEE's professional standards of membership, however, constituted a barrier to the sort of close alliance with the utilities advocated by Ferguson and Jackson. Most of the business executives whom Ferguson wanted to advise the AIEE were associates who were not eligible for full membership.[46] From about 1905 to 1912, there was mounting pressure to admit businessmen to full membership, particularly those who held engineering degrees. In 1905, the chairman of the institute's board of examiners defended the AIEE's standards. He argued that it was not the fault of the institute if a man on graduation "elects to become, for example, an author, or to adopt any other occupation where in the nature of things he cannot practice electrical engineering."[47] Although few engineering graduates had become authors, a great many had become utilities executives. President Lieb in 1905 suggested a possible compromise. While standing fast on its basic standards, Lieb thought that the AIEE might create a new grade in order to spare business executives the humiliation of being placed in the same grade with junior engineers.[48]

From 1908 to 1911, four successive committees wrestled with the problem of creating a new grade of membership. The issue was a difficult one because it went to the heart of the institute's professionalism. President Ferguson's suggestion of a close alliance with industrial leaders implied a drastic reorientation of the AIEE away from the engineering profession and toward business. On the other hand, President Lieb's proposal for a separate grade of associates would have left businessmen in a distinctly inferior position. The last committee arrived at a very ingenious compromise, which promised to make businessmen "full members," while preserving the institute's professional orientation. This was to create a new highest grade, fellow, while refashioning that of member to make businessmen eligible under certain circumstances. To preserve the professional character of the new grade member, it included not only businessmen but a large number of professional electrical engineers who were unable to qualify as full members under the old rules.

The logic of linking businessmen and junior professional engineers stemmed from the fact that electrical engineering work was controlled by a small number of highly bureaucratic corporations. For this reason many competent electrical engineers could not qualify as full members, since they did not meet the requirements for "responsible charge." By the same token, many executives were given an interest in electrical engineering because of the nature of their employment. Prior to 1901, full members had constituted one-third of the membership; with the adoption of more stringent requirements, this proportion fell, by 1905, to only 10 percent. By contrast, in the ASCE, which had similar membership standards, more than half of the membership qualified as full members. [49]

The essential element in the compromise was the professional nature of the grade of member. Engineers seeking transfer to this grade had to be able both to design and take responsible charge of electrical engineering work. But the requirements were significantly lower than for grade fellow. The candidate for member needed only to be qualified to design "under general supervision," and no time of experience at responsible charge was specified. [50] Thus, professional electrical engineers who were bureaucratic subordinates might be admitted. Although distinctly a professional grade, business executives were admitted with certain qualifications. They had to be "an executive of an electrical enterprise of large scope," with a "standing" equivalent to those who qualified under the professional requirements, and they had to be "qualified to take responsible charge of the broader features of electrical engineering involved in the work" which they directed. [51] The apparent intention of these qualifications was to restrict membership by businessmen to those who were in some sense engineers, either by education or work experience. President Dunn's explanation of the meaning of the amendments, which was sent to all members along with the proposed revisions of the constitution, stressed the "recognition of the justice of the claims of the professional element among the present Associates." He indicated that "bankers, businessmen, authors, and friends of the art of electrical engineering" would, as before, be relegated to associate membership. [52]

The formal provisions of the amendments to the constitution were probably less important than informal factors. To protect the institute's professionalism was all-important to some members. Here the crucial element was selectivity. The admission of a limited number of businessmen to a professional grade of membership would lessen discontent without compromising the AIEE's professional character. For the businessmen, the term "Member" was vital. It was the traditional name of the highest grade of American engineering societies and carried overtones implying acceptance as a "full member." In principle, both sides could achieve what they wanted most, and the amendments were carried by a record vote of the membership in May, 1912.

To reshuffle the memberhip into the new grades implied the processing of a great many applications in a short time, which would put great strain on the existing machinery of the board of examiners. In order to expedite this work, a special section to these amendments was also adopted, operative for only one year. It provided that a member or associate would "have the right" to transfer to the next higher grade, "provided he refers" to a group of fellows or members, who "upon inquiry, shall certify that he meets the requirements" of the grade. Members were required to have five certifiers to transfer to fellow, and associates had to have four certifiers to transfer to member. President Dunn, in his official statement of the meaning of this amendment, explained that applications for transfer under the special section would "not go through the Board of Examiners in the usual form of routine," and that the certifiers would act as a board of examiners. [53]

Despite the absence of any significant opposition to the constitutional amendments, they led to a bitter dispute that rocked the institute to its foundations. The controversy centered on the grade member rather than on the new grade fellow. Since all existing full members had been previously screened by the board of examiners, there was no objection to their wholesale transfer from member to fellow. The case for transfers from associate to member was quite different, since there were no professional standards prescribed for associate membership. It soon became apparent to President Mershon and others that having four members act as certifiers would not provide an adequate guarantee of an applicant's professional standing. President Mershon raised the spectre of bell-hangers and pipe-fitters becoming members and, hence, in the eyes of the public, professional engineers. [54] In actual fact, it was not technicians but businessmen who appear to have been the source of controversy. Nor was the difficulty so much the veracity of the certifiers as the ambiguity of the amendments themselves.

The precise meaning of the new requirements for the grade member was far from clear. Mershon and others who voted for the amendments assumed that they were intended to restrict the grade member to those who were, in at least some measure, technically qualified engineers. But the actual amendments were not, in fact, explicitly clear that "standing" and "responsible charge" were technical qualifications. The phrasing could equally be taken to mean bureaucratic standing and administrative responsibility. In this case, businessmen without technical qualifications, but with an interest in electrical engineering, would be eligible. Although President Mershon and other prominent members assumed that technical qualifications were necessary, a majority of the board of directors disagreed. Ex-President D. C. Jackson, an ardent spokesman for the utilities interests, insisted that executives of engineering enterprises should be considered as engineers.[55]

These two points of view came into conflict at a meeting of the board of directors in August, 1912. In form, at least, the dispute was semantic and procedural. The majority of the directors held that the amendments left them no discretion; all applications certified by the correct number of eligible members must be transferred to the next higher grade. The directors also thought that the special section suspended that part of the constitution which provided that all applications must be passed on by the board of examiners. Mershon, along with a minority of the directors, thought that applications should be referred to the board of examiners for approval. They held that the institute did have some discretionary authority in making transfers. If a candidate's record indicated that he did not meet the qualifications laid down by the constitution, the examiners might contact his certifiers and communicate their doubts. If the certifiers persisted in their endorsement, however, Mershon thought that the application should be approved anyway. But presumably some endorsers would withdraw their support if a candidate were shown to have gross deficiencies.[56]

Unable to resolve their differences, each side appealed for legal support. The board voted to refer the matter to the law committee, and Mershon requested the opinion of the institute's legal counsel. Ironically, both parties received advice upholding their opponents. The law committee agreed with Mershon that the institute did, in fact, have discretionary powers under the new amendments; and the lawyer consulted by Mershon advised him that the position taken by the majority of the directors was the correct one.[57] Neither Mershon nor the directors were in the least deterred by these adverse opinions. Underlying the legalistic arguments was a basic issue: the professional character of the institute. Mershon thought that the grade member, though no longer the institute's highest grade, implied full acceptance as a professional engineer; he was appalled at the prospect of indiscriminate mass transfers. The directors' position, although it could be justified in terms of the wording of the amendments, represented a business counterrevolution against the professional spirit of the AIEE. Large-scale transfers of businessmen to grade member would tend to orient the institute away from the engineering profession and toward industrial service. Such a change would undercut the hopes of professional unity and social action cherished by engineers like Mershon.

The refusal of either side to compromise on what was regarded as a matter of principle led to a gradual widening of the conflict. Apparently at Mershon's initiative, the board of examiners entered the fray at the next directors' meeting in September, 1912. The examiners asked that all pending applications be referred to them. When the directors refused to agree to this request, the examiners got their own independent legal opinion, one holding that referral to the examiners was mandatory under the institute's constitution. After much discussion, the directors in December offered a procedural concession. They would refer applications to the examiners, but only on the express condition that the examiners not consider "questions of professional standing" of the candidates.[58] In January the examiners rejected this compromise and requested that they be allowed to investigate the professional standing of all applicants. This the directors declined to allow, and they determined to pass on all applications for transfer themselves.

A group of prominent members attempted to mediate the dispute between Mershon and the majority of the directors. Frank J. Sprague, Michael Pupin, and Schuyler S. Wheeler, among others, urged the directors to submit the issue to a mutually acceptable lawyer for arbitration. This the board refused to do. The elder statesmen made one more effort. In February, 1913, they suggested that the interpretation of the constitution be referred to the New York Supreme Court. In this plea, they were joined by Elihu Thomson and Charles P. Steinmetz, perhaps the most prestigious members of the AIEE. But the directors were not awed by the weight of authority; they adamantly refused to budge. The directors went on to censor their own minutes to delete arguments contrary to the majority's position. Mershon opposed this action, but again the directors stood firm.[59]

The membership dispute exploded into a public controversy in March, 1913. Three of the five members of the board of examiners resigned. President Mershon, irked by the directors' censorship of the minutes, published the entire story in the AIEE *Proceedings*. Three members went even further. Dr. Louis Duncan, and Professors Michael Pupin and Francis B. Crocker, all prominently associated with Columbia University, brought suit before the Supreme Court of the State of

New York against the Board of Directors of the AIEE. They sought an injunction to restrain the directors from transferring any person "without fairly considering and deciding on merit whether such person has the qualifications required by the constitution."[60] *Electrical World* criticized these men for airing the matter in public. Cary T. Hutchinson defended the plaintiffs. "They have only one object in this action—to maintain the professional standards of the Institute," he maintained.[61]

The board of directors won a decisive victory in the AIEE membership dispute. Justice Page of the New York Supreme Court refused to grant an injunction in June, 1913. By this time the special section had expired, and the suit was abandoned.[62] The protesters had at least one other option: an appeal to the membership. They could have proposed an alternate slate of directors at the next election and a clarifying constitutional amendment. They did neither. To some extent, the elite professionals defeated themselves; many of them had long depreciated the electioneering associated with the institute's open system of nominations.[63] In 1912, an amendment was adopted which limited nominations to names suggested by a petition of fifty members.[64] It was, of course, still quite possible to place an opposition slate on the ballot, but not without an open election campaign. Not only did many members of the elite oppose this in principle, but they appear to have lacked support among rank-and-file members. The appeals of Mershon, Hutchinson, and others do not seem to have aroused much sympathy on the part of ordinary electrical engineers.

From 1913 to the middle 1920s, the AIEE underwent a gradual shift in orientation from professionalism toward business. The membership dispute was more a symptom than a cause of this change; the underlying factor was the increasing defensiveness of the electricity industry threatened by government ownership and regulation. The professional elite continued to be powerful and vocal, but they could no longer count on support for their policies. Although the drift to business control was gradual, President Mershon foresaw the danger. In his presidential address he warned against the danger that a nominating committee would put the control of the AIEE into the hands of an "oligarchy." He also depreciated the tendency of the institute to award its high offices to persons who were not active in its affairs and whose "chief claim to prominence arises from activities in fields other than that of electrical engineering as defined by the institute's constitution."[65]

At the very time that the AIEE was losing its professional momentum, a reform movement was gaining success in a rather unlikely place, the AIME. If the AIEE had been the most democratic and professional

American engineering society, the AIME had been the least active on both counts. It was ruled by its long-time secretary, Rossiter W. Raymond. Its antiquated constitution was ideally designed to perpetuate the rule of Raymond and a small clique of his New York supporters. There were no letter ballots of the entire membership; the board of directors and the secretary were routinely reelected at a sparsely attended annual meeting. Membership standards were virtually nonexistent. Raymond boasted that the AIME's membership included "common miners, laborers, mine foremen, and people that cannot spell."[66] He was even prouder to claim that 20 percent of America's most distinguished "captains of industry" were on its membership rolls.[67] But he had less to be proud of when it came to membership interest and professional achievements. There was widespread apathy among the members, and few bothered to attend meetings; by the early twentieth century the institute had declined to the level of a publishing concern.[68] "As an engineering society it is a failure," a leading technical journal commented.[69]

One cause of the dry rot in the AIME was Raymond's opposition to professionalism. He refused to print discussions of professional topics, such as unity and social responsibility. Raymond denounced the "hysteria" on the subject of ethics, and he published several articles by those opposed to written codes for engineers.[70] He banned discussions of conservation as propaganda. The AIME participated in the Conservation Congress in 1908, but Raymond helped to organize a conference the following year to denounce any conservation measures by government. A devout believer in competition and the free market, Raymond thought that the best conservation was provided by the increase of technological knowledge combined with the enlightened self-interest of businessmen.[71] For both government and engineering societies, Raymond favored a policy of strict *laissez faire*.

Although Raymond's politics antagonized many mining engineers, it was the institute's lack of membership standards that finally produced a rebellion. Led by a prominent consulting engineer, H. M. Chance, a group of members proposed to form a professional society within the AIME. Those in control of the institute received this suggestion coldly, so Chance and his friends decided to organize their own society.[72] On April 20, 1908, they founded the Mining and Metallurgical Society of America at a meeting held at the Engineer's Club in New York. The society had 115 charter members, almost all of whom were drawn from the elite of consultants, college professors, and technical journalists who were resident in New York and San Francisco. In organization and purpose, the MMSA differed radically from its parent, the AIME. The function of the new society was to represent professional mining engineers on matters of ethics and public welfare. The founders of the MMSA intended to adopt membership grades fully equal to those of the ASCE. In contrast to the highly centralized and undemocratic government of the AIME, the MMSA was to be divided into fifteen geographical districts, each of which was to elect a member of the society's governing council.[73]

Despite enthusiastic support from the technical press, the MMSA failed to live up to the hopes of its founders. The original idea was that the MMSA would function as an informal part of the AIME; membership in the institute, therefore, was to be a prerequisite to membership in the metallurgical society. But in the course of discussions leading to the creation of the MMSA, some of the founders took a position highly critical of the AIME, and at their insistence the new society was made wholly separate. This antagonism toward the AIME served to limit membership of the metallurgical society. More than one hundred of those asked to join refused to do so out of loyalty to the institute.[74]

Another liability of the MMSA was its centralization of power in New York. Lack of consideration for western members had been one of the chief grievances against the AIME, and it was second only to membership standards as a motive for forming a new society. Despite the decentralized organization of the MMSA, its New York members seem to have assumed, as a matter of course, that they should dominate the society. Virtually the only materials published by the MMSA were those contributed by New Yorkers. The New Yorkers refused to allow the San Francisco section to act independently in political matters. T. A. Rickard, the moving spirit among West Coast members, commented bitterly that "the society will suffer if . . . it is a New York affair and confined to a coterie of reformers dwelling among the sky-scrapers of Broadway."[75]

The MMSA was organized to allow mining engineers to express themselves on professional matters. But when given the opportunity, the mining engineers proved to have very little to say. There was some discussion of ethics, but no written code was promulgated. The MMSA set up a committee to recommend changes in the patent law, but no action resulted. Society members were ideologically committed to the proposition that engineers could solve social problems, but they failed to show by example how this might be done. Almost the MMSA's only positive action was to urge Congress to provide new buildings for the Bureau of Mines and Geological Survey.[76]

Although the MMSA was impotent as a separate organization, it was highly effective as a reform faction within the AIME. Its membership constituted a network of dissident engineers in mining centers across the country, virtually all of whom were professionally distinguished members of the AIME. Two events in 1911 made reform of the institute possible. The first was the retirement of Raymond, long the chief obstacle to the professional development of the AIME. The second was a proposal by the board of directors to increase revenue by creating a new class of "corporate members" who would pay higher dues. Since no professional criteria were laid down for this grade, its effect would be to make financial ability rather than standing as an engineer the test for the institute's highest grade.[77]

Opposition to the change in membership grades was organized by a group of members of the MMSA, led by C. R. Corning and G. E. Stone. They appealed for proxies from members just before the annual meeting in February, 1912. Despite the very brief notice, they were able to get 900 proxies, more than enough to dominate the meeting. The annual meetings were all important; the votes there determined elections and the adoption of constitutional amendments. They were normally lightly attended, and it had long been the custom to reelect the old directors without question. With a handful of proxies, Raymond had been able to control these meetings for years. The reformers, however, put up rival candidates, and they defeated the three regular nominees for the board of directors. "With the announcement of this vote," one engineering journal commented, "it was evident that a new *régime* had come into the management of the Institute."[78] The "insurgents" insisted that the proposed constitutional amendments be laid aside and that a committee controlled by them be created to investigate the institute's affairs. The reformers also instituted a nominating committee firmly under their control.[79]

The "stand-patters" who had long dominated the AIME were far from being ready to accept this stunning upset. The 900 proxy votes that the reformers had been able to get constituted only a small minority of the total membership. These had sufficed because the conservatives were caught off guard. Both sides, therefore, maneuvered for position for a decisive vote at the next annual meeting in February, 1913. Four major issues were debated during a year of intricate political infighting: the financial condition of the institute, the revision of the constitution, the election of a new secretary, and contests between rival slates of candidates for directors. In all of these, the progressives showed a remarkable degree of organization and political acumen. They avoided, for example, the issues of professional standards for membership and political activities. Although both were fundamental to the MMSA, to which the reformers belonged, they were matters on which rank-and-file members were less than enthusiastic. Instead the "insurgents" concentrated on their most telling popular points: financial probity and democracy in the management of AIME affairs. The conservative campaign, in contrast, was poorly organized and lacked effective leadership.[80]

On the basic issue of finances, which had triggered the revolt, the reformers had a decisive advantage. The investigatory committee found that the institute had been living beyond its means. To meet the immediate crisis the reformers cut expenses, including the pension paid to Raymond. President Kemp appealed for private donations to cover certain fixed commitments. By these means, the progressives were able to restore financial equilibrium by the end of 1912, without raising dues. For the long run, they thought that the best way to increase income would be by increasing membership. The conservatives offered no constructive program for solving the institute's financial difficulties, a fact that probably had a good deal of influence with members at the next election.[81]

Fundamental to the reformers' position was advocacy of a new constitution that would democratize the institute and prevent the recurrence of "one man rule." Led by Corning and Stone, the progressives drew up a series of constitutional amendments that incorporated drastic changes in the way the institute had traditionally been managed. The cumbersome governmental machinery was to be simplified. Proxy voting was to be abolished in favor of the more democratic letter ballot of the entire membership. The nominating committee, which had already been instituted, was to be continued as a permanent feature of the institute's government.[82] Two facts made this a "democratic" reform in the AIME context. It was vastly superior to the election system it replaced. Secondly, the AIME, because most of its members were widely dispersed, needed a mechanism for gaining consensus. The reformers were well aware of the potential abuses of a nominating committee, but they sought to offset this centralizing tendency by another reform, the organization of local sections.[83]

As with finances, the reformers' position on governmental reform of the AIME gave them a decisive advantage. The conservatives had no constructive alternative. New York control and one-man rule had long been resented; virtually no one was willing to advocate a return to the old system. The conservatives, therefore, adopted the salient points of the progressives' reforms. As a result, the new constitution was not a matter of contest; it was supported by both sides. However, the conservatives proposed a separate constitutional amendment that would create a special class of "fellows."[84] The purpose of this was not to institute professional grades of membership. Rather, the intention was to prevent a possible future merger between the MMSA and the AIME. This was, in fact, the progressives' ultimate aim, but they very wisely avoided any debate on this issue. Instead, they introduced their own separate amendments. In substance, these amendments were rather insignificant, dealing, for example, with the position of assistant-treasurer. They served, however, to stress the reformers' commitment to financial integrity. In the actual voting, the progressives' strategy was vindicated. The reformers' position, both membership control of the institute and sound finances, appealed to a wide spectrum of members, many of whom would have been less enthusiastic about membership grades and political action. The conservatives' amendment to create a special class of fellows appears to have antagonized both those favoring and those opposing membership grades.[85]

The most hotly contested issue between conservatives and progressives was the election of the secretary. The progressive-dominated nominating committee rejected the existing secretary, Joseph Struthers, who had been Raymond's hand-picked successor. They nominated Bradley Stoughton, a progressive. The conservatives nominated Struthers by petition. In one sense, the issue was purely symbolic; both sides had accepted a new constitution that would place the election of the secretary in the hands of the board of directors rather than with the membership. But as a moral issue it was central. This was a vote of confidence in the new regime. The conservatives, led by Charles Kirchoff and Karl Eilers, made the secretarial election their chief issue. They questioned the integrity of the nominating committee that rejected Struthers, and they appealed to traditional loyalty. This was a serious error. The progressives on the committee that had investigated the institute had uncovered evidence of financial irregularities by Struthers, which they now publicized. Struthers had run the annual excursions as a private venture for his own profit, making in one case as much as $5,000, although all the costs had been borne by the institute. These revelations, coming just before the election in early 1913, probably had an important affect on its outcome.[86]

Although the progressives had demonstrated political shrewdness on most issues, they blundered badly on the fourth major issue, the election of a new board of directors. The reform-dominated nominating committee went to great pains to prepare a slate of directors that would achieve geographical balance and bring able men into institute government. But the ballot they prepared was needlessly complex: voters had to select the correct number of names for one-, two-, and three-year terms; any error would invalidate the ballot. The conservatives renominated the old directors by petition. As a result, it was much easier for members to vote for the conservative directors, since they did not have to select names for particular terms. Many members found the voting instructions confusing, and few knew who the candidates were. Rather than risk spoiling their ballots, they chose the conservative directors. Thus, in the final election the progressives won overwhelmingly on all issues, except that of the new board of directors. This was to prove a very important exception.[87]

The result of the return to power of the old directors was to prevent a merger between the MMSA and the AIME. From the start, the reformers hoped to combine the two societies. This would automatically provide the institute with a professional class of members. Within a month of their victory in the February election, the insurgents opened negotiations for a merger of the MMSA with the institute. But the conservative directors refused to accept the terms demanded by the reformers. More importantly, the progressives were unable to arouse a significant amount of interest on this issue. Many members feared that the MMSA would be able to dominate the AIME, which may have been what the reformers had in mind. Ironically, another reason for the collapse of these negotiations was the success of the insurgents in revitalizing the institute. Membership was increasing, and local sections were sprouting across the country. The institute's finances were in sound condition. By their presence the reformers had given the institute a stronger professional orientation, even without formal change in membership grades. Merger with the MMSA, therefore, did not appear necessary for continued health of the AIME.[88]

Conservative control of the powerful board of directors slowed, but it did not halt, the movement for reform of the AIME. Another test of strength occurred in 1914. The progressives wanted the institute to take a positive stand on important public issues. Specifically, they wanted the institute's president to testify before Congress on a proposed revision of the mining law. Conservatives opposed as unconstitutional such a widening of the AIME's scope. The progressives countered by introducing a constitutional amendment that gave the institute the desired power. The ex-secretary, Raymond, led the opposition to this amendment. But despite his prestige, it was adopted by the members in early 1915. There were no immediate positive results, probably due to the lack of consensus among the members. But this change paved the way for later actions by the AIME in collaboration with other engineering societies. [89]

In 1916 the conservatives attempted a counterrevolution. Philip N. Moore was one of the most outstanding of the progressives; he had headed the nominating committee that unseated Struthers. The progressives wanted to nominate him for the presidency of the institute. They controlled the nominating committee, but its actions were subject to veto by the board of directors. The board overruled the committee and insisted on the nomination of a conservative, Sidney M. Jennings, for the AIME's highest office. The nominating committee complied, but suggested that the membership nominate Moore by petition. Predictably, Moore was nominated, resulting in a contested election that tested the strength of the progressives. [90]

There were two principal issues in the disputed election between Moore and Jennings in 1916. Moore's supporters complained that nominations were controlled by a small clique of New York members and that the western members were underrepresented. Jennings's followers attacked Moore for his public criticism of certain large corporations for activities contrary to the public interest. Jennings's circulars referred to his ability to ensure "the cooperation of great business enterprises" with the institute. [91] One of his journalistic allies argued that "few men not under retainer to the big corporations or otherwise in their employ are likely to have the time and opportunity to accept the responsibilities of the position." [92] The corporation issue went to the heart of the difference between progressives and conservatives. It was not so much a question of hostility toward the corporations as of the status and orientation of the engineer. Neither Moore nor the great majority of reformers were hostile to big business as such. Moore's position was that some corporations prevented engineers in their employment from acting independently in political matters. Moore, an independent consultant, symbolized the ideal of an autonomous profession. Jennings, who was vice-president of a large mining corporation, stood for a close alliance of the engineer with business. This issue, combined with western resentment of eastern control of the AIME, gave Moore a decisive victory. [93]

Moore's election encouraged progressives to make a new effort to professionalize the AIME. The fundamental issue, as always, was membership grades. President Moore advocated a very moderate and gradual change. He proposed the adoption of modest professional standards, which would apply only to new members, not to existing ones. Moore's policy of gradualism was successful. An amendment embodying his suggestions was adopted early in 1918.[94] This victory emboldened some progressives to attempt a more thorough change. They revived the proposal for a merger between the AIME and the MMSA, which would automatically create a super grade of membership substantially equal to the highest grade of the more professionally oriented societies. But the idea of austere standards failed to gain the support of rank-and-file members. Revived in the latter part of 1918, the proposal for merger was defeated at the annual meeting in February, 1919.[95]

The results of the progressive reform movements in the AIME and the AIEE were mixed. In both societies, professionals had been able to carry through important reforms. But in neither case did these initial successes lead to a sustained mass movement. Reform enthusiasm was confined to a small minority of the professional elite. They were unable to arouse lasting interest either on the part of the general public or the mass of ordinary engineers. The narrow base of their support made such reformers vulnerable to conservative counteraction, as in the AIEE after 1912. What was needed was an ideology with wider appeal than professionalism and a leader able to translate these ideas into practical programs. By 1919, the AIME was to produce such a charismatic figure in the person of Herbert Hoover. But before his emergence the initiative had shifted to the civil and mechanical engineers. Both had, by 1915, produced effective leaders. These men were able to combine the status concerns of elite engineers with the mounting economic discontent of ordinary engineers. Because of the broader ideological orientation provided by scientific management and conservation, they could link engineering reform with the larger current of national progressivism. Hoover's task, by 1919, was not to create a reform movement, but somehow to combine the more conservative and elite professionalism of the mining and electrical engineers with the radical and majoritarian mass movements that had arisen among mechanical and civil engineers. It was to be no easy task.

NOTES

1. On the "status crisis" theory of progressivism, see George E. Mowry, *The Era of Theodore Roosevelt and the Birth of Modern America* (New York, 1962), 85–105, and Richard Hofstadter, *The Age of Reform, From Bryan to F. D. R.* (New York, 1955), 131–172.

2. Harold C. Passer, *The Electrical Manufacturers, 1875–1900* (Cambridge, 1953), provides useful insights into the internal development of the electric in-

dustry. See also "Commercialized Engineering," *American Machinist*, XXX, pt. 1 (January 24, 1907), 125.

3. "Appendix. American Institute of Electrical Engineers, Rules," *Trans AIEE*, I (1884), 1.

4. "Report of Council for the Year Ending April 30th, 1894," *Trans AIEE*, XI (1894), 249. See also "Report of Council for the Year Ending April 30th, 1895," *Trans AIEE*, XII (1895), 216.

5. "Proposed Constitution of the American Institute of Electrical Engineers," *Trans AIEE*, XVIII (1901), 317–330.

6. "Annual Meeting of the American Institute of Electrical Engineers," *Electrical World*, XXXVII (May 25, 1901), 856.

7. "Report of Committee on Revision of the Rules Regarding the Election of Officers," *Trans AIEE*, IX (1892), 460–461.

8. "Annual Meeting of the American Institute of Electrical Engineers," *Electrical World*, XXI (May 27, 1893), 386, and "Institute Elections," *Electrical World*, XXIII (February 3, 1894), 138.

9. Gano Dunn, "On the Methods of Nominating and Electing Officers in the American Institute of Electrical Engineers," *Electric Journal*, IX (January, 1912), 8–12, and F. L. Hutchinson, "The American Institute of Electrical Engineers," *Electric Journal*, X (January, 1913), 23–26.

10. Charles F. Scott, "Selection of Officers for American Institute of Electrical Engineers," *Electric Journal*, VI (February, 1909), 67–68.

11. "The Presidents of Four American Engineering Societies," *Engineering News*, XLIX (January 29, 1903), 91.

12. Charles F. Scott, "The Engineer of the Twentieth Century," *Trans AIEE*, XX (July–December, 1902), 303.

13. *Ibid.*, 306.

14. *Ibid.*, 305; Charles F. Scott, "President's Address," *Trans AIEE*, XXII (1903), 13.

15. "Library Dinner of the American Institute of Electrical Engineers," *Trans AIEE*, XXI (January–June, 1903), 97–128; "Report of the Board of Directors for the Fiscal Year Ending April 30, 1903," *ibid.*, 486–496; and "A Union Building for American Engineering Societies in New York City," *Engineering News*, XLIX (May 7, 1903), 408–411.

16. "The Union Engineering Building Project," *Engineering News*, XLIX (May 28, 1903), 479–480; "Annual Convention of the American Society of Civil Engineers," *Engineering News*, XLIX (June 18, 1903), 540–542; and Charles Warren Hunt, "The Activities of the American Society of Civil Engineers During the Past Twenty-Five Years," *Trans ASCE*, LXXXII (1918), 1582.

17. "A Union Building for American Engineering Societies in New York City," *Engineering News*, XLIX (May 7, 1903), 408–409.

18. "American Institute of Electrical Engineers Affairs," *Electrical World*, LVIII (December 16, 1911), 1466.

19. Calvin W. Rice, "Joint Engineering Society Activities in the United States," *Mining and Metallurgy*, II (July, 1921), 8.

20. Charles F. Scott, "President's Address," *Trans AIEE*, XXII (1903), 13.

21. *Ibid.*, 12–13.

22. "Presentation of the Latimer Clark Library to the American Institute of Electrical Engineers," *Electrical World*, XXXVII (May 25, 1901), 865.

23. Schuyler S. Wheeler, "Engineering Honor," *Trans AIEE*, XXV (1906), 242.

24. "Discussion on 'Engineering Honor,' " *ibid.*, 266.

25. "A.I.E.E. Code of Professional Conduct," *Electrical World*, LIX (March 16, 1912), 576–577.

26. Schuyler S. Wheeler, "Engineering Honor," *Trans AIEE*, XXV (January–December, 1906), 243.

27. *Ibid.*, 245.

28. The scope of the AIEE's activities is suggested by the annual reports of the board of directors. See, for example, "Report of the Board of Directors for the Fiscal Year Ending April 30, 1913," *Trans AIEE*, XXXII, pt. 2 (May–December, 1913), 2181–82.

29. "Annual Report of the Board of Directors for the Fiscal Year Ending April 30, 1905," *Trans AIEE*, XXIV (1905), 1123.

30. "Report of the Board of Directors for the Fiscal Year Ending April 30, 1906," *Trans AIEE*, XXV (1906), 937.

31. Lewis B. Stillwell, "Conservation of Water Power," *Trans AIEE*, XXIX, pt. 2 (May–December, 1910), 1051. See also Frank G. Baum, "Water-Power Development in the National Forests, A Suggested Policy," *Trans AIEE*, XXVII, pt. 1 (January–June, 1908), 475–484.

32. Mortimer E. Cooley, "Factors Determining a Reasonable Charge for Public Utility Service," *Trans AIEE*, XXXII, pt. 2 (May–December, 1913), 2083. For another example of pro-business attitudes, see Frank G. Baum, "The Best Control of Public Utilities," *Trans AIEE*, XXXIV, pt. 1 (January–June, 1915), 145–167.

33. Steinmetz's socialism was notorious and had been the cause of his leaving Germany. For his social views, see Charles P. Steinmetz, *America and the New Epoch* (New York, 1916), *passim*.

34. Ralph D. Mershon, "Address of Mr. Ralph D. Mershon," *National Electric Light Association, Proceedings, Thirty-Fourth Convention*, II (May 29–June 2, 1911), 1021–24; "Remarks of Mr. H. L. Doherty," *ibid.*, 1024–27; and "Further Discussion by Mr. Mershon," *ibid.*, 1050–54.

35. Calvin W. Rice, "Institute Branch Meetings, Their Organization, Development and Influence," *Trans AIEE*, XXII (July–December, 1903), 63–66.

36. Charles F. Scott, "Engineering Honor and Institute Branches," *The Electric Journal*, III (July, 1906), 361–363.

37. "Report of the Board of Directors for the Fiscal Year Ending April 30, 1910," *Trans AIEE*, XXIX, pt. 2 (May–December, 1910), 1730–31.

38. Bruce Sinclair, "The Cleveland 'Radicals': Urban Engineers in the Progressive Era, 1901–1917" (seminar paper, Case Institute of Technology, 1965), *passim*. The director of Cleveland's municipally owned electric plant, Frederick W. Ballard, was secretary of the Cleveland Engineering Society; he was elected its president in 1917.

39. "American Institute of Electrical Engineers," *Electrical World*, LIX (January 27, 1912), 179–180.

40. "The Forum," *Proc AIEE*, XXXI, pt. 1 (June, 1912), 243–246.

41. "Boston Convention," *Proc AIEE*, XXXI, pt. 1 (July, 1912), 281. The directors appointed a committee to consider the sections' views, but no action resulted (see "Report of the Board of Directors for the Fiscal Year Ending April 30, 1913," *Trans AIEE*, XXXII, pt. 2 [May–December, 1913], 2170).

42. Forrest McDonald, *Insull* (Chicago, 1962), 117–118; J. Warren Stehman, *A Financial History of the American Telephone and Telegraph Company* (New York, 1967), 125.

43. Clyde L. King, ed., *The Regulation of Municipal Utilities* (New York, 1912), 253–256, 298, and Leo Sharfman, "Commission Regulation of Public Utilities: A Survey of Legislation," *The Annals, American Academy of Political and Social Science*, LIII (May, 1914), 1–18.

44. "The Convention of the American Institute of Electrical Engineers," *Engineering News*, LX (July 9, 1908), 42.

45. Dugald C. Jackson, "Electrical Engineers and the Public," *Trans AIEE*, XXX, pt. 2 (April–June, 1911), 1135–42.

46. "Annual Convention of the American Institute of Electrical Engineers," *Electrical Review*, LIII (July 4, 1908), 6.

47. "Annual Report of Board of Examiners," *Trans AIEE*, XXIV (1905), 1139.

48. John W. Lieb, "The Organization and Administration of National Engineering Societies, *ibid.*, 289.

49. "American Institute of Electrical Engineers' Affairs," *Electrical World*, LIX (January 27, 1912), 179–180, and Gano Dunn, "Notes on the Proposed Constitutional Amendments Providing for an Additional Grade of Membership," *Proc AIEE*, XXXI (February, 1912), 41–44.

50. "Parts of Constitution Affected by Proposed Additional Grade of Membership Amendment," *ibid.* 45.

51. *Ibid.*

52. Gano Dunn, "Notes on the Proposed Constitutional Amendments Providing for an Additional Grade of Membership," *ibid.*, 41, 42.

53. *Ibid.*, 42, 45.

54. "The Forum," *Proc AIEE*, XXXII, pt. 1 (April, 1913), 127, 133.

55. D. C. Jackson, "Discussion on 'A Suggestion for the Engineering Profession,' " *Trans AIEE*, XXXII, pt. 2 (May–December, 1913), 1281.

56. "The Forum," *Proc AIEE*, XXXII, pt. 1 (April, 1913), 127–128.

57. *Ibid.*, 129.

58. *Ibid.*, 130–131.

59. *Ibid.*, 131–132.

60. "Lawsuit Against the A.I.E.E.," *Electrical World*, LXI (April 26, 1913), 864.

61. Cary T. Hutchinson, "The Institute as Defendant" (letter to the editors), *Electrical World*, LXI (May 10, 1913), 995–996.

62. "Denial of Injunction Against A.I.E.E.," *Electrical World*, LXI (June 14, 1913), 1292, and "Withdrawal of Suit Against the A.I.E.E.," *Electrical World*, LXII (October 11, 1913), 724.

63. Charles F. Scott, "Selection of Officers for American Institute of Electrical Engineers," *Electric Journal*, VI (February, 1909), 67–68, and John W. Lieb, "The Organization and Administration of National Engineering Societies," *Trans AIEE*, XXIV (1905), 293.

64. Gano Dunn, "On the Methods of Nominating and Electing Officers in the American Institute of Electrical Engineers," *Electric Journal*, IX (January, 1912), 11–12, and "Nominations for Officers of A.I.E.E.," *Electrical World*, LX (December 14, 1912), 1245.

65. Ralph D. Mershon, "Some Aspects of Institute Affairs," *Trans AIEE*, XXXII, pt. 2 (May–December, 1913), 1268.

66. "Speech of Dr. Rossiter W. Raymond," in John Fritz, *The Autobiography of John Fritz* (New York, 1912), 295.

67. *Ibid.*

68. "The New Haven Meeting of the American Institute of Mining Engineers," *Engineering News*, XLVIII (October 23, 1902), 332, and "Albany Meeting of the American Institute of Mining Engineers," *Engineering News*, XLIX (February 26, 1903), 203.

69. "Mining and Metallurgical Societies–I," *Mining and Scientific Press*, XCVI (March 14, 1908), 338.

70. Rossiter W. Raymond, "Professional Ethics," *Bulletin AIME* (January, 1910), 50.

71. Rossiter W. Raymond, "The Conservation of National Resources By Legislation," *Bulletin AIME* (May, 1909), appendix, 20–36. The papers presented at this conference were published as a pamphlet (*Conservation of Natural Resources* [New York, 1909]).

72. H. M. Chance, discussion in "Meeting of Sections, New York," *Proceedings of the Mining and Metallurgical Society of America*, VI (March 31, 1913), 74–76.

73. "The Mining and Metallurgical Society of America," *The Engineering and Mining Journal*, LXXXV (April 25, 1908), 871; "The Mining and Metallurgical Society," *The Engineering and Mining Journal*, LXXXV (June 13, 1908), 1201; and H. M. Chance, "The Mining and Metallurgical Society of America" (letter to the editors), *Mining and Scientific Press*, XCVI (May 23, 1908), 698.

74. "The New Mining Society," *The Engineering and Mining Journal,* LXXXV (April 11, 1908), 773–774.

75. T. A. Rickard, "Mining and Metallurgical Society" (letter to the editors), *Mining and Scientific Press,* C (February 19, 1910), 296. See also "Mining and Metallurgical Society Problems," *Mining and Scientific Press,* CII (June 24, 1911), 838–839.

76. H. M. Chance, "The Mining and Metallurgical Society,"*Mining and Scientific Press,* CVIII (January 3, 1908), 18–19, and "Uniting the Institute and the Mining and Metallurgical Society," *Mining and Scientific Press,* CVI (March 29, 1913), 471.

77. "American Institute of Mining Engineers," *Mining and Scientific Press,* CIII (October 7, 1911), 449, and "American Institute of Mining Engineers" (letter to the editors), *Mining and Scientific Press,* CIV (February 10, 1912), 249.

78. "New York Meeting, American Institute of Mining Engineers," *Mining and Scientific Press,* CIV (March 2, 1912), 344.

79. *Ibid.* See also "Editorial," *Mining and Scientific Press,* CIV (February 24, 1912), 298, and "American Institute of Mining Engineers," *Engineering and Mining Journal,* XCIII (March 2, 1912), 438.

80. The best survey of the issues is in "To Voters of the Institute," *Mining and Scientific Press,* CVI (February 1, 1913), 203.

81. "Financing the Institute," *Mining and Scientific Press,* CV (July 20, 1912), 73; "Report of the Committee of Five," *Mining and Scientific Press,* CV (July 27, 1912), 106; "American Institute of Mining Engineers and Its Affairs," *Mining and Scientific Press,* CV (September 28, 1912), 406–408; and James Douglas, "American Institute of Mining Engineers" (letter to the editors), *Mining and Scientific Press,* CVI (January 25, 1913), 185.

82. C. R. Corning and George E. Stone, "American Institute Affairs" (letter to the editors), *Mining and Scientific Press,* CV (August 31, 1912), 278, and "Institute Constitution and By-Laws," *Mining and Scientific Press,* CV (October 19, 1912), 496–500.

83. The reformers' attitude toward a nominating committee was explained by Philip N. Moore in "Reports for the Year 1917," *Trans AIME,* LIX (February 19, 1918), xl–xlii.

84. "Constitution of the Institute," *Mining and Scientific Press,* CV (November 30, 1912), 684; "Constitution and By-Laws, American Institute of Mining Engineers," *ibid.,* 693–699; "American Institute of Mining Engineers," *Engineering and Mining Journal,* XCIV (November 9, 1912), 875–876; and "American Insti-

tute of Mining Engineers," *Engineering and Mining Journal*, XCV (January 18, 1913), 194–196.

85. "To the Voters of the Institute," *Mining and Scientific Press*, CVI (February 1, 1913), 203.

86. "Secretarial Ethics," *Mining and Scientific Press*, CVI (January 25, 1913), 168–169; "A.I.M.E. Secretaryship," *Mining and Scientific Press*, CVI (February 8, 1913), 251; and James F. Kemp, "Nominations for the Institute" (letter to the editors), *Mining and Scientific Press*, CVI (February 15, 1913), 285–286.

87. "American Institute Election," *Mining and Scientific Press*, CVI (February 8, 1913), 235, and "The Institute Election," *Mining and Scientific Press*, CVI (February 22, 1913), 301.

88. "Uniting the Institute and the Mining and Metallurgical Society," *Mining and Scientific Press*, CVI (March 29, 1913), 471; "The Institute and the Society," *Mining and Scientific Press*, CVII (September 6, 1913), 366–367; and "The Society and the Institute," *Engineering and Mining Journal*, XCVI (September 6, 1913), 466–467.

89. "Shall the A.I.M.E. Be Able to Express Itself?" *Engineering and Mining Journal*, XCVIII (December 12, 1914), 1054–56; "The Institute and Congress," *Mining and Scientific Press*, CIX (August 1, 1914), 169; Edmund G. Kirby, "Shall the Institute Express Opinions?" (letter to the editors), *Mining and Scientific Press*, CX (January 12, 1915), 31–32; and "Annual Meeting of the Institute," *Engineering and Mining Journal*, XCIX (February 20, 1915), 377.

90. "The Presidency of the Institute," *Mining and Scientific Press*, CXIV (January 13, 1917), 39–40.

91. *Ibid.*, 39.

92. *Ibid.*

93. "Mining Men in the Public Eye," *Engineering and Mining Journal*, CI (March 18, 1916), 527; H. A. Wheeler, "The Presidency of the Institute" (letter to the editors), *Engineering and Mining Journal*, CIII (February 3, 1917), 236; and "The Presidency of the Institute," *Mining and Scientific Press*, CXIV (February 24, 1917), 254–255.

94. "Editorial," *Mining and Scientific Press*, CXIV (May 19, 1917), 684; "The Affairs of the Institute," *Mining and Scientific Press*, CXVI (January 5, 1918), 3; and "Editorial," *Mining and Scientific Press*, CXVI (March 2, 1918), 283.

95. "A Merger," *Mining and Scientific Press*, CXVII (December 21, 1918), 816; "The Institute Meeting," *Mining and Scientific Press*, CXVIII (March 8, 1919), 311. Although these negotiations did not succeed, it was on this occasion that the AIME added "and Metallurgical" to its name. The official abbreviation, however, remained "A.I.M.E."

5 ▪ THE REVOLT OF THE CIVIL ENGINEERS

The restiveness of mining and electrical engineers was mild in comparison with the revolt that developed among civil engineers. The civils were more militant, in part because their economic and social position was worse. Many of the younger men were exploited. The senior men thought their social status was precarious. But another factor was that the civil engineers had a broader ideology to draw upon. Conservation provided them with the means of linking professional goals with a program of national reform. It also enabled civil engineers to unite elite professionals and junior men in a single organization, the American Association of Engineers. Because the AAE appealed to all engineers, it affected not only the ASCE, but the entire profession. Specifically, the founder societies were led to create the Engineering Council, the first major step toward professional unity and political action.

At the turn of the century, the American Society of Civil Engineers was smugly complacent. It had long since achieved the professional goals that electrical and mining engineers were struggling to gain. It had austere entrance requirements that excluded all but professional engineers from its higher ranks of membership. It embodied, at least in theory, the principle of professional unity; civil engineers were all engineers not engaged in military service. Its members, therefore, rejected Carnegie's offer in 1903 of a unified headquarters for the four major engineering societies. They viewed with disdain the lower entrance requirements of other societies and saw the proposed building as a scheme by the other societies to bask in the ASCE's accumulated prestige. Those civil engineers who had been able to secure for themselves an occupational role as independent consultants looked down on the employed engineers in other societies as unworthy of being considered truly professional.[1]

But behind the placid exterior of the ASCE, the profession was in ferment. The society was living on the prestige of the past, when civil engineering had been the queen of applied sciences. Those days were over. Much of the glamour of engineering had passed on to the newer fields, such as electrical engineering. But the schools continued to turn out great numbers of civil engineers, who found that the profession was overcrowded. Economic pressures, especially on the younger civil engineers, added to a pervasive discontent in the early twentieth century. Other factors influenced even those engineers who were financially comfortable. Set in a pattern of an older age, the ASCE gave little support to the new aspirations of engineers. It did little to support engineers in government service or to encourage a wider application of engineering to social problems. Above all, the society did nothing to assuage the mounting status anxieties of American engineers. It refused, for example, to adopt a code of ethics—a step advocated for its prestige value, rather than for any serious concern with self-policing. It rejected unity with other societies. It showed no interest in a policy of militant professionalism, at a time when drastic action appeared to many engineers to be the only course that might salvage the status of their profession.

"We see, then, that the businessman is the master; the engineer is his good slave," commented one engineer.[2] This statement, published in *Engineering News* in 1904, precipitated a flurry of letters and editorials on the subject of the status of the engineer. Two points emerged from this discussion with stark clarity: the engineer was poorly rewarded and lacked power in relation to his employers. Some engineers thought the only remedy lay in the engineer becoming a businessman. Others vehemently dissented. The engineer was a scientific man and should pursue his profession even if it did not make him wealthy, they argued.[3] The debate over the role and status of the engineer was intensified by an editorial in the New York *Evening Post*, in 1907, entitled "Commercialized Engineering." The editors pointed out that the alliance of engineers with predatory monopolies was contrary to the interests of the public as well as the profession. They went on to suggest that professional societies should determine "what the relations of special talent should be to employing corporations, what the obligations of a man to the severe demands of science when they are in conflict with mere money-making."[4] *Engineering News* condemned commercialism and urged engineering societies to adopt codes of ethics to check the "business view" of engineering.[5] Charles Whiting Baker, the editor, summed up the progressive viewpoint in a signed article. "And is the engineer a mere passive looker on in this struggle to establish anew our liberties?" he asked. He suggested that the engineer was more than a "cog in the industrial machine," and that by vigorous action he could establish his rightful position as "more than a faithful slave, or a faithful overseer."[6]

Of all civil engineers, it was the younger men who were the most alienated by the ASCE's conservatism. They had virtually no say in the society's policies, since junior men could not vote in society elections. The society was geared to the interests of an elite of senior and success-ful men; it showed no real concern for the plight of the younger men. Indeed, the senior men sometimes exploited the younger engineers. As one younger engineer bitterly commented, "we can expect very little encouragement from the generals of the profession who are mainly responsible for the conditions in which young engineers find them-selves."[7]

In the spring of 1909, the young men took matters into their own hands by forming a protest organization, the Technical League. Its membership, confined to civil engineers, was at first secret, since its leaders feared retaliation on the part of employers. Largely centered in New York, the Technical League's principal aim was the "closing" of the engineering profession. This they proposed to achieve by a licensing law that would restrict the practice of engineering to those with four years of college education. Other aims included publicity, the inculca-tion of professional ethics, and the elevation of the dignity and *esprit de corps* of civil engineers.[8] The president of the Technical League denounced the ASCE as "a mere private corporation ruled by a hand-ful, mostly New Yorkers."[9]

The new organization met hostility and criticism from older engi-neers. President Onward Bates of the ASCE, while acknowledging that the engineer's lot was not satisfactory, insisted, "there is no room in the profession for trades unionism."[10] A. L. Dabney noted that he used to employ uneducated persons as rodmen and chainmen, but that he had found college-trained engineers were willing to accept such jobs for the same compensation. He criticized the Technical League on the ground that there would be less demand for young engineers at higher prices.[11] Charles Whiting Baker expressed a more sophisticated opposi-tion, which was typical of the attitude of many progressive engineers. He blamed the overrapid expansion of engineering education for the overcrowding of the profession. He thought the remedy lay in broaden-ing the training of engineers, presumably by lengthening the course of study. This would serve the twofold purpose of limiting the supply of engineers and preparing engineers to serve as leaders of public affairs.[12]

The young men were not content to wait for a long-run solution. The Technical League got a licensing law introduced before the New York Legislature in 1910. To get a license, a candidate would be required to have four years of college. The bill was intended to cover all engineers, subordinates as well as those in responsible charge of work.[13] This measure met the united opposition of senior civil engineers, progressive and conservative alike. A number of prominent engineers testified against the bill, including the president of the ASCE, John A. Bensel. J. L. Raldiris, the sponsor of the licensing law, suggested a compromise: that both parties agree on a model law.[14] The ASCE went along reluctantly with this suggestion. In 1910, a special committee of the society on licensing helped to draft the model code. The ASCE's bill watered down the Technical League's proposals. This measure did not close the engineering profession; it applied only to senior men and it did not require a college degree.[15]

The Technical League appears to have collapsed soon after the failure of its licensing bill. But its very existence and its attraction of approximately one thousand members in New York suggested that discontent among younger engineers was widespread. It had a permanent impact in changing the ASCE's policy of isolation. The ASCE's special committee on licensing concluded the society should assist in the formulation of all laws that affected civil engineering.[16] In the following year, 1911, the ASCE joined with the other founder societies in their Joint Conference Committee. Originally concerned with the Engineering Societies Building, this committee became a spokesman for the founder societies on matters of public policy. Its initial work seems to have been chiefly opposition to licensing laws sponsored by the Technical League. But by 1915, it had become involved in advising the convention drafting a new constitution for New York. The committee concerned itself with such issues as the tenure of public officials, a suggested department of public works, and the regulation of public utilities.[17]

The younger engineers were not the only ones discontented with the ASCE's policies; progressives were unhappy because of the lack of genuine democracy within the society. Only members who held the two highest grades could vote at all. In 1900, the society created a nominating committee controlled by the board of directors.[18] Contests for society offices were rare. The society opposed, at least informally, any publicity of its internal affairs. One officer who violated this taboo was defeated at the next election.[19] The absence of publicity made it difficult for the average member to form an opinion of what was going on, which was probably more important in limiting membership control than any formal regulations.

Pressure for the reform of the ASCE came from a rather loose alliance of progressive civil engineers. Their most vocal spokesman was probably Charles Whiting Baker, editor of *Engineering News*, the leading technical journal in civil engineering. The progressives' greatest strength lay outside of New York, where ASCE members sometimes felt left out by the dominant group. Many such members were active in local engineering societies. These organizations had lower entrance requirements than the ASCE, but they were often more involved in advancing professionalism than was the national society. The reformers, therefore, favored a drastic decentralization of the ASCE, with the creation of autonomous sections that would cooperate in professional matters with local societies. They hoped to make these sections the basic building blocks of the society: each holding its own elections and governing its own affairs. Above all, progressive engineers wanted their profession to act positively to advance the welfare and status of engineers; they rejected the "fallacious idea" that the only functions of engineering societies were holding meetings, discussing papers, and producing technical literature. But progressives insisted that before organizations like the ASCE could undertake new responsibilities, they should become "truly representative of the membership at large."[20]

From 1909 to 1914, progressive reformers attempted to democratize the governmental machinery of the ASCE. In 1909, Mr. Samuel Whinery proposed a resolution that denounced the nominating committee as "unrepresentative and unsatisfactory."[21] But he received little support. By 1913, the directors were willing to make a small concession: to increase the number of districts from seven to thirteen. This fell short of the radical decentralization demanded by Gardner S. Williams and other reformers. Progressives urged the need for a code of ethics, the formation of an employment bureau, and further decentralization and democratization of the society.[22]

In the resulting battle over the constitution, the progressives were decisively defeated. Three amendments were proposed for the vote of the membership in 1914. The board backed an amendment to increase the number of districts from seven to thirteen. Progressive members introduced two additional amendments. The first of these would have drastically decentralized the society. After a first nominating ballot, the members of the nominating committee would be elected by the letter ballot of the district they represented. A second amendment would have curbed the authority of the secretary, Charles W. Hunt, by excluding him from the board of directors and reducing his powers.[23] This amendment was probably a serious blunder. Hunt, who had been secretary since 1895, was honest and able. He had many supporters within the society. A circular in support of Hunt and opposed to the reform amendments was signed by 732 members.[24] When the final ballots were counted in 1914, all of the amendments were defeated; none got a majority of the society's voting members, much less the two-thirds necessary to amend the constitution.[25]

Although victorious, the conservatives were aware of the necessity of meeting some of the demands of the discontented. The leaders of the ASCE were willing to offer three substantial concessions to the dissidents. First, the society in 1914 adopted a code of ethics.[26] Second, the ASCE abandoned its outdated claim to represent the entire profession, and it began to cooperate with other major engineering societies. After preliminary discussions in 1915, the ASCE voted, in 1916, to join the other founder societies in the Engineering Societies Building in New York.[27] Finally, the society accepted a significant measure of decentralization. Progressives had organized a number of local sections on an extralegal basis. Within a short time, twenty-one of these associations were in existence. In 1915, the society accorded a measure of recognition to them by holding a meeting of the presidents of these local organizations at society headquarters. Another step was the formal division of the society into thirteen districts by a constitutional amendment adopted in 1915.[28]

Although the progressives were somewhat mollified by the concessions offered them, the younger men remained as discontented as ever. The code of ethics, like those previously adopted by the ASME and AIEE, treated engineers as if they were consultants. It had little relevance to the problems of younger, employed engineers.[29] The society continued to deny that any real economic problem existed. In 1913, the ASCE set up a committee to investigate the salaries of civil engineers. The report appeared to indicate that civil engineers were fairly well-off economically. But critics pointed out that the committee used the membership of the ASCE as a sample, although the society was, by design, an elite organization that included only one-twelfth of all civil engineers. In any case, the committee had received replies from only one-half of those circularized.[30] A second report in 1915, based on a larger sample of ASCE members, gave similar results, however, and indicated that some senior men were receiving very large earnings. *Engineering News* criticized these results on the ground that most of the well-to-do members got their money by engaging in business rather than by professional activities.[31] Whatever their true meaning, these reports provided the society with justifications for doing nothing to aid the younger men.

The progressives were less discontented than the younger men, but they were freer to act. Many were men of considerable professional stature; they did not have to fear the pressure of employers. Moreover, the progressives had at hand a natural avenue for expressing themselves, the local engineering societies. Several of the informal associations of ASCE members entered into cooperative arrangements with local societies to advance mutual professional interests.[32] The membership of these societies was predominantly made up of civil engineers. They had no inhibitions about an active role in politics. Perhaps the most active and successful was the Cleveland Engineering Society.

The Cleveland Engineering Society was the most progressive of all the local engineering societies. Founded in 1880 as a club for civil engineers, the Cleveland society resented the pretensions of the ASCE to speak for the entire profession. To local pride there was added a strong progressive current, induced in large part by the urban reforms initiated by Mayor Tom L. Johnson and continued by his successor, Newton D. Baker. The Cleveland Engineering Society developed close ties with both Johnson and Baker, as well as with local civic organizations. Although the Cleveland society lacked the high membership standards of the ASCE—three years of experience in engineering or a related field was all that was necessary—the Clevelanders showed a professional spirit that greatly exceeded that of the torpid ASCE. From 1904 to 1910, the Cleveland Engineering Society developed institutional means for bringing the engineers' knowledge to bear on local problems. It established a network of committees staffed by experts in such fields as franchises, municipal ownership, water supply, rapid transit, smoke abatement, and legislation affecting engineering. Thus, it was ready to advise public officials on almost any civic matter involving engineering.[33]

The Cleveland Engineering Society was concerned as much with advancing professionalism as with civic problems. Like other engineers, those in Cleveland were deeply concerned about the status of their profession; they thought that the engineers' work was inadequately understood by the general public. Unlike other engineers who merely bemoaned this situation, the Clevelanders acted. In 1912, the society created a publicity committee with Charles E. Drayer as its chairman. Drayer asked local editors why the newspapers gave so little coverage to the activities of engineers. From them he learned the form of a news release and other rudiments of public relations. Drayer showed a genuine flare for this work, and his material appeared frequently in local newspapers. He went on to sponsor a series of articles on "Engineering as a Life Work," which appeared in the *Cleveland Plain Dealer* and, later, as a book.[34] Drayer's publicity attracted national attention among engineers. His work was discussed at length in the pages of *Engineering News* and before the ASME. Drayer was not content with only a local reputation; he attempted to persuade other engineering societies to undertake similar work.[35]

Like most progressive engineers, Drayer's horizons were narrowly limited; his skills at organization and publicity were relevant to the status anxieties of senior men. They failed to arouse the interest of younger engineers. Professionalism was not enough; a major overhaul of civil engineering would require the efforts of both progressives and younger engineers. To achieve this unity, a broader program and a more inclusive ideology were needed. The conservation movement provided civil engineers with just such a program and set of ideas. Conservation was a euphemism for scientific planning applied to natural resources. It could be expanded into a campaign for the engineering of society generally. It combined a stress on democracy with a commitment to professional values, which gave it the potentiality of uniting all of the dissident civil engineers in a single reform crusade.

Conservation, as an ideology, was eminently suited for use by engineers. Like the engineers' thinking, conservationist rhetoric alternated between the poles of "is" and "ought." Conservation borrowed from American democratic thought. It could be presented as a battle of the people against the interests, of the little man against selfish monopolists. But it was also a campaign for scientific planning by experts.[36] It was this very ambiguity that made conservation so useful for engineers. Its democratic appeal allowed the uniting of disparate engineering factions and the linking of the aspirations of engineers to a general national reform movement with wide popular support. At the same time, its stress on planning permitted its use to advance professionalism. Effective planning required autonomy for the expert; a large expansion of planning would thrust engineers into a position of social leadership that would greatly enhance their status. It would bring material benefits that would ameliorate the lot of the younger engineer.

Conservation had at least one drawback as an ideology: it was by no means primarily an engineers' movement. It had been founded by a geologist, Major John Wesley Powell, and its chief spokesman in the progressive era was a forester, Gifford Pinchot. But an important number of engineers were involved, especially in building dams and irrigation works. Although overshadowed by Pinchot, Frederick Haynes Newell, the founder and first director of the Reclamation Service, was an important national leader of the conservation movement. Engineers might lay claim to conservation, since it was unquestionably applied science. Its very inclusiveness had advantages. Civil, mining, and electrical engineers were all involved in it professionally. As an example of science applied to society, it appealed to all engineers.

Prior to 1915, Newell and other conservation leaders had not become involved in the attempts to reform the ASCE. On the contrary, Newell and his followers had cooperated with the conservative oligarchy controlling the ASCE. They had, for example, opposed the proposal that the society join in the common headquarters building in 1903.[37] They had worked to defeat the progressives' attempt to reform the ASCE in 1913 and 1914.[38] But this alliance was more a matter of tactics than conviction. Newell and other conservationists were unhappy over the fact that they got no active support from their profession in their battles within the federal bureaucracy. Newell thought the ASCE, along with the other founder societies, actively discouraged their members from attempting a wider application of engineering to society.[39]

Newell was convinced that the scope of conservation should be greatly expanded. He thought the disinterested viewpoint of the engineer would be as fruitful when applied to problems of transportation, communication, public utilities, agriculture, housing, and education as it had proven to be in the field of natural resources. "In all of these matters, which pertain to the conservation and use of the resources of the country, both material and human, and the development of ideals," Newell maintained, "the engineer should be the leader."[40] The past triumphs of science and technology inspired him with the hope that the same methods would enable man to master human affairs.[41] Newell assumed that scientific laws governed society as well as the material world. Before a congress on "human engineering," Newell argued that there were laws in industry as certain as those in nature, and that by discovering and applying these laws, engineers could pull "humanity from this slough of discord."[42]

Newell recognized that the engineering of society implied treating man as a material. "Primarily, under our new conception of things," Newell held, "the engineer is concerned with the greatest of all the forces used in engineering, that of man himself."[43] The engineer should influence society on two levels: by the design of social units and by the control of men's thinking. Human groups, Newell thought, should be examined by engineers as if they were machines "in which the wheels and bearings are men and not metals."[44]   In addition, Newell wanted engineers to undertake "the beneficial control of human forces and sentiments."[45]

The Orwellian implications of Newell's philosophy were obscured by his habit of clothing his thought in moralistic rhetoric. Thus, Newell and other engineers usually avoided such terms as "planning," preferring such euphemisms as "conservation," the "elimination of waste," and "service to society." He saw conservation as a moral drama; it was a struggle between the "aggressive minority" intent on personal gain and the reformer full of "altruistic ideals."[46] The role of the engineer, Newell maintained, should be of a "missionary of light and progress."[47] Newell thought of the engineer as a "pioneer of a better and higher degree of civilization."[48] He discussed proposals for lobbying and propaganda by engineers in terms of "social responsibility" and "cooperation." Although Newell's objectives quite explicitly included the improvement and elevation of the status of the engineer, he insisted that the engineers' true motives were altruistic.

Conservation, like other aspects of American reform thought at the turn of the century, looked both forward and backward. Newell shared the progressive's Janus-like posture. He saw conservation as the germ of a new scientifically planned society. But the aim was to preserve traditional individualistic virtues threatened by urban, industrialized America. By means of reclamation, Newell hoped to reopen the frontier and revive the values of pioneering hardihood and self-reliance. Reclamation would provide homes for farm families, thus strengthening the values of rural life against the encroaching city. He disliked and distrusted the sprawling industrial cities and the immigrant peoples who flocked there; by scientific planning, traditional values could be preserved, and American economic and social institutions defended. This, rather than economic considerations, provided the ultimate justification of conservation.[49]

Newell's vision of engineering planning was closely linked to his belief in professional autonomy. His stress on the independence of the engineer was one reason his ideas of planning were so authoritarian. To Newell, autonomy was the essence of professionalism. He defined the professional engineer as one who was "working independently, directing his own affairs with the maximum of personal freedom."[50] In contrast to the engineer, Newell maintained, the doctor or lawyer "is called in *not* to carry out instructions of an employer, but on the contrary to dictate to the man who ultimately pays the fee. . . . He ceases to be a professional man the moment he takes orders from an employer."[51]

Newell, as director of the Reclamation Service, sought to preserve his independence by adroit and far-sighted administrative practices. He kept complete records, including a personal diary, with which to defend himself. His motto was, "in time of tranquility prepare for investigations."[52] With the assistance of Senator Newlands and other congressional friends, Newell attempted to obtain the maximum possible independence for his agency.[53]

It was, in part, Newell's idea of professional independence that led to his dismissal as head of the Reclamation Service in December, 1914. Newell wanted the settlers to pay the full cost of irrigation works. These funds could then be used to finance more reclamation projects. In this way Newell would gain a measure of fiscal autonomy. But the westerners complained that the works were too expensive, and they refused to pay. Newell was able to meet these and other challenges to his authority so long as he had the support of Theodore Roosevelt. But under Roosevelt's successors, Taft and Wilson, Newell was at odds with successive secretaries of the interior. Several matters of policy were involved, but Newell's own stubborn independence was at least one factor in his troubles. Newell's power in the Reclamation Service was steadily whittled down until his final dismissal.[54]

Newell's firing was not too surprising in the light of Pinchot's earlier and more spectacular separation from government service. But more significant were the differing responses of the forester and the engineer. Pinchot reacted by throwing himself into the "Bull Moose" third-party movement, and he continued to be a fixture of American reform politics through the New Deal period. Newell directed his attention to a narrower audience, his own profession. He attempted to unite the engineers, raise their status, and direct their efforts to social reform.

From personal experience, Newell was convinced that the founder societies would never unite or act effectively. He compared them with outdated fossils. They were dominated by business interests and were consequently unfit to lay down high professional standards. At the same time, their exclusive membership standards left out many engineers. A professional organization dedicated to professional improvement and social reform would have to be democratically governed and broadly inclusive in membership.[55]

On February 8, 1915, Newell recorded in his diary, "New idea on engineering federation. Wrote to Charles Whiting Baker."[56] This new idea was to dominate Newell's thought for the next six years and was destined to shake the engineering profession to its foundations. In furthering this project, which he christened "engineering cooperation,"

Newell received vital support from Baker and *Engineering News*. In essence, Newell envisaged by-passing the founder societies by creating a federation of local and regional engineering societies. There would be a national headquarters, a full-time secretary, and a president. But most of the professional activity would be organized locally. The president was to travel across the country, coordinating the efforts of the individual societies.[57] Newell attempted to line up support from leading engineers, and he made some progress. One of his most valuable converts was C. E. Drayer, who proved to be ideal second-in-command. The alliance of Newell and Drayer marked the beginning of a more militant phase in the progressivism of civil engineers.

The idea of engineering cooperation set forth in numerous articles and speeches by Newell and Drayer centered around a four-point program. Engineers should unite—or cooperate—in order to create an employment bureau, improve professional ethics, secure the passage of better laws, and carry out publicity. Newell was concerned that the wealthiest "engineers" who dominated the founder societies were not true professionals: they practiced engineering as a business. He hoped by a strong code of ethics to check commercialism and the power of this class in professional affairs. Publicity, such as Drayer had organized for the Cleveland Engineering Society, was fundamental as a means of raising the status of engineers. Newell, like many engineers, assumed that social status and position were determined by esteem and respect accorded by the general public. The importance of the engineer's work in modern civilization justified a higher status than he enjoyed; if only the public were properly informed of the engineer's merits, he would be accorded the deference that was his just due. Better laws implied not merely civic helpfulness, on the pattern of the Cleveland Engineering Society, but a major reorganization of American society. Newell thought of the impending restructuring of society as an inevitable development, like the rise of science and technology; and he saw in the wartime planning measures of the major belligerents during World War I a foreshadowing of the new order. "Out of the white heat of the devouring conflict in Europe," he predicted, "a new world is emerging."[58] This world would be planned; the question was, who should do the planning. He thought that the engineers were better qualified and that the task should not be left "to the commercial interests, to the bankers, to the lawyers, or to the politicians." To draw attention to these aims, Newell and Drayer published and spoke extensively on "cooperation." To secure positive action they organized a series of conferences on engineering cooperation.

The first cooperation conference was held at Buffalo, New York, concurrently with the annual meeting of the ASME in June, 1915. Newell and Drayer were apparently attempting to capitalize on the discontent within the ASME organized by a group of dissident young mechanical engineers, whose leader was Morris L. Cooke. Newell had contacted Cooke and apparently obtained some support, and Cooke arranged for Drayer to give a paper on his publicity work in Cleveland. But the attempt to use a national society as a vehicle for sponsoring a movement based on local societies was probably a mistake. Cooke and the other mechanical engineers hoped to work through the ASME; they did not favor by-passing the national societies. Their orientation was toward scientific management, rather than conservation, and they differed with Newell on a number of important issues. The meeting was not a success for Newell. Attendance was small, and the delegates recommended against forming a new organization "at this time."[59]

Newell recovered quickly from this initial setback. He broadened the base of his movement by creating a Committee on Engineering Cooperation, with himself as president and Drayer as secretary, to which a number of prominent engineers lent their names. With the assistance of Baker, Newell and Drayer engaged in a large-scale publicity campaign to pave the way for a second conference on cooperation. Newell drew up a circular letter, which he distributed widely. He traveled extensively and had talks with a large number of prominent engineers. Newell's thorough preparation paid off at the second conference on cooperation, which was held in Chicago on April 13, 1916. This time the conference was attended by delegates from forty-two local and regional engineering societies, and great enthusiasm was engendered. Although the conference did not take any immediate action, it endorsed several resolutions. The most important, which began "let us have a central organization," authorized Newell to select four other engineers, who would be empowered to draw up a plan for the unification of the entire profession. Other resolutions followed Newell's proposals: the creation of an employment bureau, support of engineers in public service, and periodic conferences of delegates from member societies to discuss matters of general interest.[60]

At the third conference, which met on March 29 and 30, 1917, Newell presented the plan of unification drawn up by his subcommittee. It called for immediate unification of the engineering profession, with or without the founder societies. This proposal led to a floor fight. A prominent member of the ASCE, Gardner S. Williams, introduced an alternate resolution, which would have prevented action by the local soci-

eties. Williams reported that the founder societies had agreed to form
their own unity agency, the Engineering Council. Williams's resolution
merely requested the founder societies to expand their proposed council
to permit membership by other engineering societies—local, state, and
national. The delegate of the Cleveland Engineering Society led the
opposition, asserting that "neither the Cleveland Engineering Society
nor the members of the ASCE in Cleveland are willing to place their
interest in this enterprise in the hands of the national societies."[61] Most
of the delegates, however, thought that support by the founder societies
was essential for any unity organization, and Williams's resolution was
carried. Newell's cooperation movement appeared to be a total wreck.

At this point Newell demonstrated his remarkable versatility and
ingenuity. He abandoned the Committee on Cooperation, but found a
new means to the same end in a protest organization of young engineers, the American Association of Engineers. Since the Technical
League had faded away, there had been no channel through which the
younger civil engineers could express their grievances. A number of
proposals circulated among the younger men, including the idea of
forming a labor union. Finally, in 1914, a group of civil engineers in
municipal employment in Chicago formed the Associated Technical
Men. This organization soon split into two factions, one favoring and
one opposing the formation of a labor union. The antiunion engineers
approached a number of prominent engineers, including Newell, to
enlist their support in the formation of a new society, the American
Association of Engineers.[62] The motives of the older engineers, who
were able to exercise a controlling influence in the AAE from its inception, were mixed; although sympathetic with the young men, they also
feared their radicalism. One reason that Newell and other senior men
joined the AAE was to guide the new organization into safe channels.
Above all they wanted the AAE to avoid such "labor-union" tactics as
licensing or collective bargaining.[63]

Initially Newell had no thought of using the AAE as a vehicle for
unifying the engineering profession. The organization of the AAE took
place just as Newell was preparing for his second cooperation conference; and until the failure of the third conference, in 1917, Newell preferred to work through the local societies. The AAE was made up of
young men, most of whom did not belong to any engineering society,
and most of whom did not yet rank as "professional" engineers. Like the
other senior men, he wished to head off "selfish" policies, particularly a
licensing law such as the Technical League had advocated. At the first
meeting of the AAE in Chicago in September, 1915, Newell assured the

younger men that it would not be necessary to restrict entry into the profession, since there would be a great increase in the demand for engineers once the public was made aware of their potential usefulness in solving social problems. At the same time, Newell warned that no organization could survive if it was founded on purely selfish motives. Newell saw no direct connection between the AAE and his own cooperation movement; he argued that they had separate, though harmonious, objectives.[64] But after the failure of the third conference on cooperation, Newell and Drayer transferred their attention to the AAE. Aided by the fact that the AAE's early leaders were young men who were drafted or volunteered for war service, Newell and Drayer took over the association. In 1918, Drayer became secretary of the AAE, and Newell was elected president the following year.

The results of Newell's and Drayer's leadership of the AAE were spectacular. Between January, 1919, and September, 1920, the association increased its membership from 2,300 to 20,000.[65] Underlying the rapid increase in membership was a growing militance among younger engineers. Engineering salaries tended to lag far behind the postwar inflationary spiral, and the economic pinch on younger engineers intensified. Newell and Drayer were very active in promoting the AAE in their travels across the country. But the most important factor probably was a drastic change of policy on the part of both Newell and the AAE. A. H. Krom, who was one of the founders and the man responsible for bringing Newell and other senior engineers into the organization, had followed a conservative policy. As secretary of the AAE through 1917, Krom favored a policy of educating the public and "salesmanship."[66] He opposed licensing and collective bargaining. Under Newell's and Drayer's leadership, the AAE, in 1919, adopted licensing as its fundamental tool for improving the position of engineers.[67] It lobbied vigorously for licensing laws for engineers in several states, and within a year it claimed success in both New York and Virginia. The AAE took some tentative steps toward collective bargaining. In 1919, the association drew up salary schedules for engineers employed by the railroads and presented these to the federal mediating machinery, and in this way was able to get substantial wage increases for some of its members. The AAE, however, denied that it was a labor union, and it did in fact forswear the strike and other union tactics.[68] But the AAE's new militance did much to swell its membership rolls over the 20,000 mark.

While compromising on tactics, Newell did not lose sight of his fundamental objective: the unification of the engineering profession through a federation of local societies. Newell and Drayer were able to remodel the AAE to approximate the unity organization they had planned in 1915. At the end of Newell's year as president, he assumed the newly created position of "director of field forces," a sort of traveling chief executive. This was a unique office for an American engineering society, but it had been sketched originally by Newell, in 1915, as an integral part of his proposed unity organization. More significant for Newell's revised conception of the AAE was the idea of working out reciprocity agreements between the AAE and local and regional engineering societies. By an exchange of membership the local society would become, in effect, a chapter of the AAE. In August of 1918, Drayer reported the first such agreement, with the Cleveland Engineering Society.[69] Similarly, Newell transferred to the AAE his hopes for the engineering of society. He urged local chapters to study engineering applied to social problems. "The same genius which has enabled the engineer to control the floods and direct electric energy," Newell told the members of the AAE, "should enable him to study effectively and turn to the benefit of humanity the great forces wrongly employed or lying latent in human needs and desires."[70]

Newell's cooperation movement and, even more, the AAE were threats to the traditional leadership of the founder societies. The ASCE was the society most endangered, since more than two-thirds of the members of the AAE were civil engineers.[71] One civil engineer commented, "if we cannot recruit into our organization the younger men, we are going to be lost in ten years."[72] By 1918, the discontent of the younger men had become a serious issue for all of the founder societies. J. Parke Channing, a mining engineer, warned the members of the ASCE that the older men were in danger of being overshadowed by Newell's vigorous leadership of the young. While believing in democracy "as a whole," Channing thought that more experienced engineers would be better able to deal with public affairs.[73]

The founder societies' answer to Newell's challenge was the Engineering Council. Founded in April, 1917, the council was a means of acting for the entire profession in nontechnical matters. During its two and one-half years of existence, the council's chief function was to forestall the AAE. President George H. Pegram of the ASCE thought that the profession should assist the younger engineers, but that the council should act with a "conservatism commensurate with its dignity and responsibility."[74] In practice, this meant that the council adopted watered-down versions of many of the reforms pioneered by the AAE. It drew up a model licensing law. This it used to combat the more radical proposals of the AAE. By 1921, nineteen states had adopted licensing laws for engineers.[75] The council undertook publicity work for engineers, it set up an employment bureau, and it adopted a set of salary schedules for engineers. The council claimed to have secured the reinstatement of 350 engineers in New York whose positions had been in danger.[76]

The Engineering Council was less successful in speaking for the profession in matters of public policy. This reflected, in part, the essentially negative motives of its founders. The council was so structured as to make positive action very difficult. Its membership was initially restricted to the four founder societies, each of which had a veto. The council functioned as a committee of ambassadors; representatives had to vote according to the instructions of the boards of directors of their societies. In effect, this system of voting gave a veto to any group strong enough to influence the board of any one of the four founder societies. Many industrial interests had this much power.[77]

The Engineering Council expressed a conservative interpretation of the engineer's social role and responsibilities, one which did not distinguish the engineer from the businessman. In 1919, the council joined the National Chamber of Commerce. A press release by the council noted that "in joining this great association of trade and commercial organizations, Engineering Council adds its strength to organized business."[78] Before joining, the council first had to endorse the official creed of the Chamber of Commerce. This included a strong statement of *laissez faire* philosophy. The government should stay out of the fields of transportation, communication, industry, and commerce.[79] That this endorsement was seriously considered was shown by the council's stand on water power. The council opposed government development of hydroelectric power, and it followed a line similar to that of the electric utilities. The council expressed the hope that the opposition to a "good" water-power bill, which was led by "Wisconsin senators, will be broken through the influence of Wisconsin engineers."[80]

On only one issue did the Engineering Council succeed in arousing the enthusiasm of a sizeable number of engineers. The council revived the idea of creating a cabinet-level Department of Public Works to be headed by an engineer. The council called a national conference in April, 1919, to discuss this issue. It was attended by seventy-four engineering societies. They created the National Public Works Department Association to agitate for a law. The association, which was financed by the council, drafted a bill and was able to get it introduced in Congress. Although these activities got favorable attention in the technical press, they had little impact on the public at large or on Congress.[81]

That the Engineering Council did perform some useful services was due in large part to the election of a progressive mining engineer, J. Parke Channing, as president. In its first year the council had no budget at all, and under these circumstances its activities were few and limited. When Channing was elected in early 1918, he obtained an appropriation of $16,000 for 1918, which he managed to get increased to $23,000 in 1919. An ardent enthusiast for the cooperation of engineers, Channing contributed $25,000 of his own money to enable the council to open a Washington office. By means of this office and a network of special committees, the council was able to perform a diversity of services to the profession.[82]

Probably the most important achievements of the council were in advising the government during the First World War. The council helped the army and navy find specialized engineering talent. Through a special committee, the council assisted the military in examining some 135,000 suggested inventions. The council advised the Fuel Administration and the Bureau of Mines on fuel conservation and other problems. Although relatively unspectacular, these activities were of considerable value.[83]

Despite Channing's best efforts, the council was unable to carry its wartime momentum into the peace. The council set up a committee on public affairs, whose chief function appears to have been to steer the council "away from questionable or unprofitable undertakings."[84] The council took no stand on the many great issues of the day. It did not propose, as many engineers were urging, a great national program of reconstruction based on engineering principles. It did not call for a crusade against waste. It confined its attention to rather minor issues. The council helped cause the unification of fourteen government mapmaking agencies into the Board of Surveys and Maps. It concerned itself with the reform of the patent office. It urged the appointment of engineers to government positions. These actions notably failed to arouse the interest of either rank-and-file engineers or the general public.[85]

The Engineering Council was a failure. It was unable to articulate an engineering viewpoint on postwar problems. It failed to win the support of the engineering profession. Newell's supporters attacked the council as unrepresentative and undemocratic.[86] Their charges were not without foundation. Even within the founder societies the council aroused little enthusiasm. Other engineering societies were even less attracted. Channing attempted to widen the council's base of support during the war through the device of the War Committee of Technical Societies. In this way, the cooperation of some seven national engineering societies was secured.[87] But the council's attempt to recruit new members after the war was not successful. Of some eighty eligible societies, only two joined, the American Electric Railway Association and the American Society for Testing Materials.[88] Both were more trade associations than professional organizations. Most of all, from the point of view of its founders, the Engineering Council failed to halt the growth of the AAE, which experienced its greatest increases in membership in the years 1917–1920.

If the Engineering Council failed to unite the engineering profession, so also did the AAE. The association's chief appeal was to young civil engineers. Despite Newell's best efforts, the AAE accomplished even less than the Engineering Council in the area of social responsibility. Newell was unable to wean the younger men from their preoccupation with their own material interests. In any case, it was doubtful whether such a decentralized organization could have attacked national problems effectively. A more flexible body of doctrine and an abler group of leaders were to be found among the mechanical engineers, particularly among those devoted to scientific management.

NOTES

1. "Annual Convention of the American Society of Civil Engineers," *Engineering News*, XLIX (June 18, 1903), 540–542; "Letters Regarding the Union Engineering Society House," *Engineering News*, XLIX (May 21, 1903), 455–456; and J. J. R. Croes, "The Union Engineering Building Project" (letter to the editors), *Engineering News*, XLIX (May 28, 1903), 479.

2. J. Swinburne, "The Engineer and the Business Man," *Engineering News*, LII (December 22, 1904), 568.

3. "The Engineer and the Business Man," *Engineering News*, LIII (January 5, 1905), 20–21.

4. "Commercialized Engineering," *Engineering News*, LVII (January 17, 1907), 75.

5. "Engineering and the Commercial View," *Engineering News*, LVIII (October 24, 1907), 446–447.

6. Charles Whiting Baker, "The Engineer as a Man Among Men," *Engineering News*, LIX (February 20, 1908), 197–198.

7. Young Engineer, "The Engineer's Compensation and His Attitude Toward the Profession" (letter to the editors), *Engineering News*, LXI (May 20, 1909), 553.

8. Executive Committee of Technical League, "Improvement in the Economic Status of the Engineer" (letter to the editors), *Engineering News*, LXI (June 17, 1909), 665–666.

9. W. G. Eliot, "Making a Profession of Engineering" (letter to the editors), *Engineering News*, LXV (May 11, 1911), 575.

10. Onward Bates, "Address at the Forty-First Annual Convention," *Trans ASCE*, LXIV (September, 1909), 574.

11. A. L. Dabney, "Rates of Pay for Subordinate Engineering Assistants," *Engineering News*, LXII (July 1, 1909), 19.

12. Charles Whiting Baker, "A Broader Training for the Engineer," *Engineering News*, LXI (May 20, 1909), 536–538.

13. *Engineering News*, LXIII (March 24, 1910), 343.

14. Edward L. Raldiris, "The Raldiris Bill Requiring a State License for Civil Engineers Practicing in New York," *Engineering News*, LXIII (April 14, 1910), 438.

15. "Annual Meeting of the American Society of Civil Engineers," *Engineering News*, LXV (January 26, 1911), 113, and "A Law Requiring Engineers to be Examined and Licensed," *ibid.*, 106–107.

16. "The Engineering Society as a Molder of Legislation," *Engineering News*, LXV (June 30, 1910), 758–759.

17. "A Committee of Engineers Recommends Amendments to the New York Constitution," *Engineering News*, LXXIII (April 29, 1915), 842, and Calvin Rice, "Joint Engineering Society Activities in the United States," *Mining and Metallurgy*, II (July, 1921), 8.

18. "Minutes of Meeting of the Society, October 3, 1900," *Proc ASCE*, XXVI (October, 1900), 215–216.

19. "Contests in Engineering Society Elections," *Engineering News*, LXXII (December 31, 1914), 1319–20.

20. "What Can the Engineering Profession do to Improve Its Position?" *Engineering News*, LXI (August 17, 1911), 211.

21. "Minutes of Meetings of the Society, Fifty-Sixth Annual Meeting," *Proc ASCE*, XXXV (February, 1909), 49, 66–69.

22. "Report in Full of the Sixtieth Annual Meeting," *Proc ASCE*, XXXIX (February, 1913), 80.

23. "Report in Full of the First and Second Sessions and of the Business Meeting of the Forty-Fifth Annual Convention," *Proc ASCE*, XXXIX (August, 1913), 430–443.

24. *Ibid.*, 443–449.

25. "Proposed Amendments to the Constitution of the American Society of Civil Engineers Defeated," *Engineering News*, LXXI (March 12, 1914), 589.

26. "American Society of Civil Engineers in Convention Approves Code of Ethics," *Engineering News*, LXXI (June 4, 1914), 1275.

27. "Society of Civil Engineers Votes for Co-Operation," *Engineering News*, LXXV (June 29, 1916), 1241.

28. "Changes in the Organization of the National Engineering Societies," *Engineering News* LXVI (August 3, 1911), 148–149; Charles Warren Hunt, "The Activities of the American Society of Civil Engineers During the Past Twenty-Five Years," *Trans ASCE*, LXXXII (1918), 1596, 1600.

29. "Proposed Code of Ethics of the American Society of Civil Engineers," *Engineering News*, LXXI (April 30, 1914), 987.

30. "Conditions of Employment and Compensation of Civil Engineers," *Engineering News*, LXXI (January 22, 1914), 166, and "The Compensation of Engineers," *ibid.*, 199–201.

31. "The Compensation of Civil Engineers," *Engineering News*, LXXIII (January 7, 1915), 34–37.

32. "The Importance of the Local Engineering Society," *Engineering News*, LXXII (December 31, 1914), 1320–21, and "Co-Operation or Amalgamation of Local Engineering Societies," *Engineering News*, LXXIV (November 25, 1915), 1042.

33. Bruce Sinclair, "The Cleveland 'Radicals': Urban Engineers in the Progressive Era, 1901–1917" (seminar paper, Case Institute of Technology, 1965), 1–5, *passim*.

34. *Ibid.*, 20–23; C. E. Drayer, "The Publicity Work of the Cleveland Engineering Society," *Engineering News*, LXXI (January 22, 1914), 176–179; F. H. Newell and C. E. Drayer, eds., *Engineering as a Career* (New York, 1916).

35. C. E. Drayer, "The Engineer and Publicity," *Journal ASME*, XXXVII (February, 1915), 88–92; "Publicity Work by Engineering Societies," *Engineering News*, LXIX (May, 1913) 1134; and "Annual Reports, The Publicity Committee," *Journal of the Cleveland Engineering Society*, VII (July, 1914), 87–88.

36. Samuel P. Hays, *Conservation and the Gospel of Efficiency* (Cambridge, 1959), 1–5, *passim*.

37. "Opinions of Washington Members of the American Society of Civil Engineers," *Engineering News*, LI (February 11, 1904), 131. Newell himself took no active part in this and other battles, but A. P. Davis, his chief lieutenant, spoke for the engineers in the Bureau of Reclamation.

38. "Report in Full of the First and Second Sessions and of the Business Meeting of the Forty-Fifth Annual Convention," *Proc ASCE*, XXXIX (August, 1913), 447.

39. F. H. Newell, "The Engineer in the Public Service," *Engineering News*, LXVIII (July 25, 1912), 153, and F. H. Newell, "Ethics of the Engineering Profession," *The Annals of the American Academy of Political and Social Science*, CI (May, 1922), 84–85.

40. Engineer [F. H. Newell], "Engineering Profession Should Have Leading Part in Reconstruction," *Engineering News-Record*, LXXXI (October 17, 1918), 730. For Newell's authorship, see University of Illinois, *Books and Articles Published by the Corps of Instruction, May 1, 1918–April 30, 1919* (n.p., n.d.), 26.

41. F. H. Newell, "Address of Frederick H. Newell as Retiring President of the American Association of Engineers," *Professional Engineer*, V (June, 1920), 10.

42. F. H. Newell, "The New Emphasis on the Human Factor in Industry," *Ohio State University Bulletin*, XXI (January, 1917), 104–105.

43. Engineer [F. H. Newell], "Engineering Profession Should Have Leading Part in Reconstruction," *Engineering News-Record*, LXXXI (October 17, 1918), 729.

44. F. H. Newell, "The Engineer in Public Service" (ms. of address, May, 1912), p. 13, notebook "Newell 1912," box 6, Papers of Frederick H. Newell, Manuscripts Division, Library of Congress, Washington, D.C.

45. F. H. Newell, "The Engineer as a Citizen," *Journal ASME*, XXXVII (July, 1915), vi.

46. F. H. Newell, *Water Resources: Present and Future Uses* (New Haven, 1920), 30.

47. F. H. Newell, "Awakening of the Engineer," *Engineering News*, LXXIV (September 16, 1915), 568.

48. F. H. Newell, "The Engineer in Public Service" (ms. of address, May, 1912), p. 6, notebook "Newell 1912," box 6, Newell Papers.

49. F. H. Newell, "Progress in Reclamation of Arid Lands in the Western United States," *Annual Report of the Smithsonian Institution . . . June 30, 1910* (Washington, 1911), 171; F. H. Newell, "National Efforts at Home Making," *Annual Report of the Smithsonian Institution . . . June 30, 1922* (Washington, 1924), 517, and F. H. Newell, "The Human Side of Engineering," *Engineering Record*, LXX (August 29, 1914), 236.

50. F. H. Newell, "A Practical Plan of Engineering Co-Operation," *Journal of the Cleveland Engineering Society*, IX (March, 1917), 311.

51. *Ibid.*

52. F. H. Newell, "The Engineer in the Public Service," *Engineering News*, LXVIII (July 25, 1912), 154.

53. Hays, *Conservation and the Gospel of Efficiency*, 135–136, 156.

54. *Ibid.*, 245–248.

55. F. H. Newell, "Ethics of the Engineering Profession," *The Annals of the American Academy of Political and Social Science*, CI (May, 1920), 78–85, and F. H. Newell, "A Practical Plan of Engineering Co-Operation," *Journal of the Cleveland Engineering Society*, IX (March, 1917), 308–317.

56. Diary of Frederick Haynes Newell, February 8, 1915, box 2, Newell Papers.

57. "Engineers Will Discuss Plans for Nation-Wide Cooperative Movement," *Engineering Record*, LXXI (June 19, 1915), 766–767.

58. F. H. Newell, "A Practical Plan of Engineering Co-Operation," *Journal of the Cleveland Engineering Society*, IX (March, 1917), 308.

59. "Co-operation of Engineers in Publicity Work," *Engineering News*, LXXIV (July 1, 1915), 41.

60. Diary of Frederick Haynes Newell, April 13, 1916, Newell Papers; "Conference on Engineering Co-operation at Chicago," *Engineering News*, LXXV (April 20, 1916), 771–772; and "Report of Committee on Engineering Cooperation," *The American Society of Heating and Ventilating Engineers, Transactions*, XXII (1916), 408–415.

61. "Proceedings of the Third Conference of Delegates From Engineering Societies Called by the Committee on Engineering Co-Operation, Chicago, March 29–30, 1917," p. 93, *passim*, file "Eng. Coop. Conference," box 5, Newell Papers. See also Gardner S. Williams, "Engineering Co-operation Outside the National Societies," *Bulletin of the Federated American Engineering Societies*, II (April, 1923), 6.

62. "Employment of Engineers," *The Monad*, I (July–August, 1916), 5–9; Diary of Frederick Haynes Newell, August 24, 1915, Newell Papers.

63. Isham Randolph, "Glad Question Three Was Defeated," *Engineering News-Record*, LXXXIV (May 13, 1920), 978, and Gardner S. Williams, "Engineers National Cooperative Business Organization," *The Monad*, II (June, 1917), 14–15.

64. F. H. Newell, "Awakening of the Engineer," *Engineering News*, LXXIV (September 16, 1915), 568–570.

65. "Growth of A.A.E. to 20,000," *Professional Engineer*, V (September, 1920), 22.

66. A. H. Krom, "Organization for Engineers," *The Monad*, III (February, 1918), 7–9, and A. H. Krom, "Selling A.A.E. Memberships to Engineers," *The Monad*, I (February, 1916), 52–55.

67. "Change in License Policy," *The Monad*, IV (November, 1919), 19.

68. "First Big, Worthwhile Step in Compensation Efforts," *Engineering News-Record*, LXXXII (March 27, 1919), 597; C. E. Drayer, "The Present Activities and Future Responsibilities of A.A.E.," *Engineering News-Record*, LXXXIV (May 20, 1920), 1023–24; and F. H. Newell, "The American Association of Engineers," *Journal of the Engineers' Club of St. Louis*, V (January–March, 1920), 32–39.

69. C. E. Drayer, "Reciprocity Between National and Local Societies," *Engineering News-Record*, LXXXI (August 1, 1918), 217–218.

70. F. H. Newell, "Address of Frederick H. Newell as Retiring President of the American Association of Engineers," *Professional Engineer*, V (June, 1920), 10.

71. "Official Report of the National Cooperative Convention," *The Monad*, III (June, 1918), 8–34.

72. "The Sixty-Seventh Annual Meeting," *Proc ASCE*, XLVI (February, 1920), 195.

73. "Letter from Chairman Channing," *Proc ASCE*, XLV (October–December, 1919), 933.

74. George H. Pegram, "Address at the Annual Meeting," *Trans ASCE*, LXXXII (1918), 165.

75. "Defeated Bill for Licensing Engineers to be Fought Over in Massachusetts," *Mining and Metallurgy*, II (December, 1921), 25–26; E. W. Ellis and others, "Licensing and Registration of Engineers in the United States," *Mining and Metallurgy*, XXVI (January, 1945), 22.

76. "Extracts from Annual Report," *Mechanical Engineering*, XL (April, 1919), 408.

77. J. Parke Channing, "Engineering Council Enters Larger Sphere," *Mining and Metallurgy*, I (December, 1920), 17–18, and "Engineering Council and Non-Member Societies," *Engineering News-Record*, LXXXII (May 1, 1919), 888–889.

78. "Engineering Council," *Mechanical Engineering*, XLI (November, 1919), 910.

79. Engineering Council Minutes, June 6, 1919, pp. 7–8, American Engineering Council Papers, Engineering Societies Library, New York.

80. "Engineering Council," *Mechanical Engineering*, XLI (November, 1919), 912.

81. J. Parke Channing and others, *Engineering Council, A Brief History* (New York, 1921), 6, and "Suspension of National Public Works Department Association," financial statement, file N, box 1-I/584, Herbert Hoover Papers, Herbert Hoover Presidential Library, West Branch, Iowa.

82. Alfred D. Flinn, "History and Future of Engineering Council," *Mining and Metallurgy*, I (December, 1920), 15–16.

83. *Ibid.*

84. Channing, *Engineering Council*, 6–8.

85. *Ibid.*, 5.

86. C. E. Drayer, "Will the Engineering Council Satisfy the Demand?" *Engineering News–Record*, LXXVIII (May 3, 1917), 277–278, and "Engineering Council is Challenged," *Engineering News-Record*, LXXVIII (May 10, 1917), 321.

87. R. N. Inglis "The Engineering Council's War Committee of Technical Societies," *Mechanical Engineering*, XL (May, 1918), 402–403.

88. Channing, *Engineering Council*, 3–4.

6 ▪ MEASURING THE UNMEASURABLE:

SCIENTIFIC MANAGEMENT AND REFORM

Frederick W. Taylor, the inventor of scientific management, played a unique role in the development of the social thought of engineers. Scientific management was an attempt to realize in practice the ideas that other engineers only talked about. Taylor thought that he had discovered the scientific laws of management. By means of them, he thought he could engineer certain crucially important areas of society and thus attack social problems such as national efficiency and class conflict. He showed in a concrete manner how engineers might resolve their status dilemmas. He reconstructed certain industrial bureaucracies in such a way as to enhance professionalism. He demonstrated that scientific management opened up a new field for independent consulting practice. Taylor's followers pioneered a larger social role that they thought would ultimately make engineers leaders of society, increase their power, and enhance the deference accorded them. Through an impressive body of doctrine and a highly developed technology, the Taylorites offered engineers a means for the fulfillment of their professional aspirations.

Not least among the virtues of scientific management was that it forced engineers to face explicitly problems that were latent in the profession's thinking. Although the ambiguities and paradoxes in the engineers' thought were many, two were crucial. First, engineers assumed that human society was governed by scientific laws; by discovering and applying them the engineer could solve social problems. Second, engineers thought that by undertaking a larger social role they could resolve their own status dilemmas. The Taylorites were no better equipped philosophically to deal with these questions than were other engineers, but they were forced to do so by the very nature of their work. Taylor and his followers had to determine the precise nature of the laws assumed to apply to man. They were compelled to define the proper position of an autonomous professional man within an industrial bureaucracy and the role of an expert planner in a democratic society. If it cannot be said that Taylor and his followers found the correct answers, they at least asked some of the right questions.

Taylor, a scion of an aristocratic Philadelphia family, seemed destined to follow a well-worn groove from Exeter Academy to a genteel law practice. Taylor, however, abandoned jurisprudence and opted to become a mechanical engineer. He prepared himself for this new career by means of apprenticeship in a Philadelphia machine shop. In 1878, he was employed by the Midvale Steel Company, where he became the protégé of William Sellers, a Germantown neighbor of the Taylor family. It was Sellers who was responsible for Taylor's spectacular rise in six years from workman to shop superintendent, and it was he who supported the experiments in metal cutting and time study at Midvale that were the foundations of Taylor's later fame.[1]

Taylor's initiation was not exceptional. Upperclass Philadelphians had been educating their sons in this manner for some time. It was a recognized means of recruitment for elite mechanical engineers. In this way Taylor received not only an excellent training but a strong professional orientation as well. He supplemented these by correspondence study, leading to a degree in mechanical engineering from Stevens Institute of Technology.[2] Although Taylor's career owed something to family connections, he quickly demonstrated his genius. Inarticulate with words, Taylor's virtuosity in dealing with mechanical matters made him one of the greatest American engineers of all time. To him invention was a dissipation that had to be curbed. Had he never devised scientific management, his discovery of high-speed steel, his work in metal cutting, and his many other engineering innovations would have sufficed to make him famous.[3]

Several factors induced Taylor to apply his genius to the problems of management. One was the changing context of mechanical-engineering practice. The mechanical engineer of the old school typically became the head of his own machine shop and thus enjoyed the social status accorded to independent businessmen. But Taylor and others of his generation were employed by large corporations, and their chances of becoming entrepreneurs were limited. This dilemma, typical of modern engineering, has often been resolved by the engineer identifying himself with business. But another alternative was open, which Taylor followed: to assimilate management within the engineering profession. This implied studying management like a field of engineering in order to discover the underlying scientific laws assumed to govern it. In essence, this is what Taylor's scientific management was all about.[4]

A second reason for Taylor's interest in management was the labor problem. At Midvale, Taylor encountered the workers' systematic limitation of production, which he called "soldiering." Taylor was deeply offended. It was not just a matter of efficiency; to him soldiering was a great moral evil. "Unfortunately for the character of the workman," Taylor maintained, soldiering involved deception, which led to a further deterioration of morals. The workers became intemperate, they lacked thrift, and their initiative declined. The result was that "the employer is soon looked upon as an antagonist, if not as an enemy."[5] Mutual confidence and enthusiasm disappeared. Soldiering appeared to Taylor to be the crux of the labor problem. By attacking it he would get at such fundamentals as the erosion of traditional values, the alienation of the workers, and class conflict.

Taylor's first response to soldiering by the men at Midvale was simply to force them to work harder. This led to something bordering on warfare between Taylor and his men. After three years of struggle, Taylor succeeded in raising output, but found this no recompense for the bitterness his policies aroused. Taylor therefore sought a new approach that would make the interests of the workmen and management the same. He turned to science for a restatement of the old doctrine of the harmony of interests. If he could discover the scientific laws underlying management, then he could eliminate both the soldiering of the workers and the arbitrary authority of management. In his initial efforts, Taylor attempted to find the law governing a full day's work by seeking a correlation between fatigue and the number of foot-pounds of work performed. Such a straightforward mechanical solution eluded him after several trials, but he never completely gave up hope.[6]

At Midvale, Taylor laid the foundations of scientific management. He built upon them in the course of further work as a management consultant with a number of firms, reaching a climax at the Bethlehem Steel Company from 1898 to 1901. Thereafter Taylor retired from active practice in order to promote his system of management; the actual work of application he left to a growing body of disciples. Henry L. Gantt developed the task-and-bonus system of wage payment, an improvement on Taylor's own differential piece rate. Carl Barth, the mathematician of the group, invented a new type of slide rule for determining the optimum speeds and settings of machinery. Frank Gilbreth extended Taylorism to the building trades and devised new methods of time study. Morris L. Cooke, one of the last to join the movement, applied scientific management to the printing industry. But his greatest interest and achievements lay in publicity and the application of Taylor's principles to politics. None of these later modifications altered the fundamentals of Taylor's system. Among the most important of these basic principles were an incentive wage, time study, a planning department, functional management, and the exception principle.[7]

Taylor's point of departure was an incentive system for paying wages. In this he was not alone. The movement of mechanical engineers into large industry produced a spate of papers before the ASME on the labor problem. Most of these proposed the linking of wages to productivity. By increasing his efficiency, the worker would raise both his wages and the profits of management. However, Taylor's approach differed from that of other mechanical engineers. Where they left the speed of the work to the discretion of the men, Taylor attempted to calculate exactly how much work should be done in a given time. To do this he had to replace guesswork and rule-of-thumb by exact knowledge for every step in production.[8]

It was altogether fitting that the popular symbol of Taylor's system came to be the stop watch; for time study was the heart of scientific management. As developed by Taylor, this involved two parts, analysis and synthesis. In analysis, the engineer divided a particular job into its elementary movements. He discarded the nonessential ones and carefully examined each of the remainder to determine the quickest and least wasteful means of performing the operation. As a preliminary to synthesis, the engineer then described, recorded, and indexed these elementary motions, along with the percentages required to cover necessary delays. In the second stage, synthesis, the engineer combined elementary operations in the correct sequence to determine the time and the exact method for performing any given task. This information was presented to the worker in the form of a card with written instructions that left nothing to the worker's own discretion.[9]

Time study and its implementation led, of necessity, to a series of fundamental changes in industrial organization. To assign daily tasks and to prepare the written instructions required a planning room or department. It would not be enough, for example, simply to give the worker instructions as to what to do for a particular job. The sequence of machines used and the route followed by the materials also had to be specified. Another necessity was complete standardization of machines, tools, and all plant operations. Without this, the specific times required for a task would be meaningless. Moreover, the work of a large number of people must be coordinated. This was especially necessary as Taylor insisted on an elaborate division of labor; each separate subtask was assigned to a man trained for that specific job. To bring these together smoothly involved elaborate, systematic planning. The planning department became the nerve center for the new system of management. "The shop, and indeed the whole works," Taylor held, "should be managed, not by the manager, superintendent, or foreman, but by the planning department."[10]

Taylor's reorganization of the pattern of bureaucracy and, by that token, of authority, went much deeper than simply adding one new department to an existing structure. Taken together, Taylor's reforms amounted to a sweeping reorganization. Most profoundly, he wanted to change the very basis of bureaucratic authority. He proposed to make skill and knowledge the basis of power. This led to a change in bureaucratic organization.[11] Taylor rejected the traditional linear hierarchy or "military" line of command. In its place Taylor substituted functional management. Each foreman or manager would be responsible for only a single function, such as quality control, speed of work, maintenance, or discipline. Each separate function of management would be grouped into its own hierarchy. Thus, the various speed bosses would report to the speed foreman, and so on. The usual military chain of command would be replaced by a series of parallel hierarchies based on skill. Although developed on the shop level, Taylor insisted that "throughout the whole field of management the military type of organization should be abandoned, and what may be called the 'functional type' substituted in its place."[12]

In one sense, functional organization simply extended to management Taylor's emphasis on a more elaborate division of labor and on the need for special training and expert knowledge in the performance of each task. Only in this way could rule-of-thumb and guesswork be replaced by exact knowledge. But the fuller implications of functional management became apparent when it was combined with another of Taylor's doctrines, the "exception" principle. At each level of management, the responsible person should receive information only about exceptions to routine, not everyday performance. A higher echelon would be called upon to decide and act only when there was a breakdown in the plan of operations. This would limit the authority of the top layers of management. Executives would not concern themselves with the details of matters they did not understand, but would confine themselves to those things about which they did have adequate knowledge and about which they were competent to make decisions.[13]

The manifest function of Taylor's restructuring of the industrial bureaucracy was to increase efficiency. But it would also perform the latent function of enhancing the autonomy of engineers. The stress throughout on exact knowledge would bolster the position of all experts. Time study would transfer the skills of the worker to the engineer. This would provide a basis in esoteric knowledge for the classical defense of professional autonomy. Through the planning department, engineers would administer the ordinary operations of the entire plant. Their situation would be further improved by functional management and the exception principle. By the functional system experts would be judged only by other experts. The exception principle would grant the maximum of independence to each layer in the hierarchy; top management would take virtually no hand in directing the day-to-day activities of lower echelons.

Taylor's proposed reconstruction of industry would benefit the elite of the profession even more than the mass of employed engineers. Each separate productive unit would require conscious, rational design. This would open up a vast new field for consulting practice in management engineering. Taylor himself pioneered this new role, and he trained a group of followers who practiced his methods as independent consultants. Installing scientific management was a lengthy process, requiring from two to four years. During this period Taylor insisted it was absolutely necessary that the engineer responsible be given complete authority or the attempt would fail. This demand constituted one of the chief impediments to the acceptance of scientific management by business, but Taylor stubbornly refused to budge. Taylor had opened the possibility of an independent role for engineers in an area in which their position had been that of bureaucratic subordinates. He was not willing to compromise this newly found autonomy.[14]

Although the Taylor system offered obvious benefits to engineers, it threatened the power of both labor and traditional management. Workers would lose their skills, and these would be transferred to management. They would no longer determine the pace at which they worked, and they would be subject to a more severe discipline. Management would have to accept the complete authority, even dictation, of the engineer at the time the new methods of management were installed. When the system was in operation, the scope of traditional management would be greatly curtailed.

Taylor was well aware of the difficulties in persuading management and the workers to accept limits on their freedom. He, therefore, called for a "mental revolution" on the part of both sides. He regarded this revolution as the very essence of scientific management. It required that both parties lay aside their traditional antagonism and cooperate to further the common task of production. To make this possible, Taylor proposed to substitute law for force. "Both sides must recognize as essential the substitution of exact scientific investigation and knowledge for the old individual judgement or opinion, either of the workman or the boss," Taylor held.[15] He could boast that under his system all autocratic authority was eliminated. In theory, at least, the engineer, too, was subordinate to this law. But, in practice, there was this difference: the engineer was the lawgiver. Although Taylor did not stress the point, engineers had long been arguing that they should serve as arbiters in settling the strife between labor and capital, and in this way enhance their status. But if his system was to enable engineers to play this role, Taylor had to insist that the system was truly scientific.[16]

Taylor's insistence on the scientific nature of his management discoveries was fundamental. Taylor's work could, and ultimately did, stand on its own as a collection of efficiency devices. As such it might be a useful tool kit for management. But Taylor wanted it to do more than that. Scientific management can best be seen as an extension and codification of the engineers' ideology. Central to this formulation was the idea of a universe composed of overarching laws of nature. The engineer saw himself as the discoverer and interpreter of these laws. They provided the basis in esoteric knowledge for a defense of autonomy and other professional values. Only if Taylor's discoveries were truly scientific would they fulfill those professional ideals. Only if these laws were moral as well as material could they allow engineers to serve as the judges of what were admittedly moral issues. Thus, in meeting the social responsibilities of his profession, the engineer would become a leader and reformer of society, and in this way he would greatly enhance his own status. Taylorism carried the professional aspirations of engineers from dream to concrete reality, at least in the eyes of its founders. But unless the new management rested on absolute, scientific laws, it was but one managerial system among many.

Taylor wanted the ASME to endorse the scientific claims of his system of management. A committee was set up to examine the question. To the secretary, Leon P. Alford, Taylor wrote that his system consti-

1 MEASURING THE UNMEASURABLE

tuted a "new type of management, in which scientific laws and rules are substituted for the old-fashioned individual initiative of the men."[17] But he failed to convince the committee. It held that management was an art that involved using the laws of chemistry and physics.[18] Taylor was deeply disappointed. This was particularly the case as the chairman, James M. Dodge, was an old friend. Dodge rejected Taylor's scientific pretensions on two grounds. Taylor had used rather arbitrary definitions of the "average" man and "first-class" man in standardizing his results. Also, as Dodge pointed out, rather large percentages had to be added to the calculated time for a job to allow for unavoidable delays. Determining these amounts was also arbitrary. Taylor admitted that there were "one or two" elements that were not based on exact knowledge, but he asserted that this was also true of "all other sciences." He thought that future investigations would eliminate these gaps. Taylor concluded with an affirmation of his faith: that in the end his system would reduce all the elements of industry to "exact science."[19]

Taylor was convinced that the scientific laws he had discovered were moral as well as material in character. By identifying the good with mechanical efficiency, he blurred the distinction between "is" and "ought." Taylor thought that time study determined not only how quickly a piece of work could be done, but how fast it ought to be done. Thus, he thought he could measure precisely a fair day's work.[20] Similarly, by experimentation with various incentives, he had discovered how much he had to pay the workers to get them to work to their capacity. This allowed him, or so he thought, to calculate a fair day's pay.[21] From a just wage and a proper day's work, it was possible to go further. Indeed, if Taylor was correct, he had laid the foundations for a scientific ethics.

At the time Taylor formulated his system, there was a widespread belief that science might determine ethics.[22] It has persisted, with diminishing vigor, well into the twentieth century. Social Darwinism was perhaps the foremost example of a "scientific" doctrine that proposed to solve ethical questions, and it was the chief immediate inspiration of American engineers, including Taylor. The idea of a scientific ethics, however, was never universally accepted. One reason for its declining popularity in recent times was its rejection by the scientific community. This was part of a larger movement to divorce science from metaphysics. The critical analyses of the foundations of science by such men as Ernst Mach, Pierre Duhem, and Henri Poincaré had a profound influence on the leaders of twentieth century physical science. Poincaré's rejection of a scientific ethics was perhaps the simplest and most straightforward. He wrote:

> If the premises of a syllogism are both in the indicative, the conclusion will also be in the indicative. For the conclusion to have been stated in the imperative, at least one of the premises must itself have been in the imperative. But scientific principles and geometric postulates are and can be only in the indicative. Experimental truths are again in that same mood, and at the basis of the sciences, there is and there can be nothing else. [23]

Taylor was attempting to measure the unmeasurable.

Taylor's conviction that he had in fact discovered the key to a scientific ethics was shown by his relations with Scudder Klyce, a naval engineer and amateur philosopher. After hearing Taylor lecture, Klyce became convinced that Taylor had "succeeded in establishing a *complete* and comprehensible standard of ethics." [24] In Klyce's formulation, scientific management was linked to social Darwinism. He held that there were laws governing the universe, and that the most fundamental of these were the preservation of the race and the need to care for others. Together these two constituted "the whole moral law." In Klyce's view, Taylor's great achievement was to have quantified this moral law in terms of "dollars and cents." Taylor's system of paying the workers constituted a "fair and just" measure of the cooperation required by society. [25]

Taylor received Klyce's ideas with great enthusiasm. He had copies of Klyce's long, somewhat incoherent, letters typed out and distributed among his followers and friends. Taylor wrote that Klyce "seems to me to state the principles of scientific management very clearly," and he though his ideas "remarkably sound." [26] He offered Klyce a position of leadership in the scientific management movement in developing the "true fundamental theory." [27] Taylor's interest in Klyce was all the more remarkable in that Klyce showed signs of mental derangement; at the time he wrote to Taylor he was going through one of a series of nervous breakdowns. Klyce's ideas on science were extremely eccentric. In his first letter to Taylor, Klyce had argued that the "whole of thermodynamics is a very poor approximation," which he thought "hardly worthy to be called a real science." [28] Klyce also thought that he had demonstrated the falsity of the laws of mechanics and the inadequacy of the law of gravity. Despite these unorthodox views, Taylor helped prepare a digest of Klyce's ideas, which he got published in *The Outlook* under the title "Scientific Management and the Moral Law."[29] There can be little doubt that Taylor shared Klyce's opinions of the ethical significance of scientific management.

Taylor's belief that he had, in fact, discovered the laws of a new exact science was manifested on many occasions; it ran like a thread through all of his writings. He assumed that his laws had the same characteristics other scientific laws were thought to possess. Taylor was convinced that his laws, like those of physics, reflected an underlying reality. In physics this substrate consisted of atoms moving in Euclidian space. Taylor thought that he had discovered the "atoms" of work in "elementary motions." These were the things measured by time study. By adding them together, Taylor could determine the time and motions for any job. This idea was extended by one of Taylor's followers, Frank Gilbreth, who employed motion picture cameras and other devices to analyze these elementary motions in greater detail.[30]

In the latter nineteenth century, it was generally assumed that scientific laws were both universal and deterministic. Taylor thought that scientific management had these characteristics also. If the laws of management were universal, they applied not just to factories, but to all social organizations. Taylor insisted that

> the same principles can be applied with equal force to all social activities: to the management of our homes; the management of our farms; the management of the business of our tradesmen, large and small; of our churches, our philanthropic institutions, our universities, and our governmental departments.[31]

These laws were assumed by the Taylorites to be deterministic; that is, from a given set of initial conditions a single, predictable result would follow. This idea found expression in scientific management in the idea that any given managerial situation would have a unique solution, "the One Best Way." It was the task of the systematizer to find this "One Best Way."[32]

The universality of scientific management made it a suitable instrument for the reform of all manner of social institutions. Even without its metaphysical encrustations, scientific management had a large reform potential. Taylor called for a campaign for national efficiency comparable to the conservation movement.[33] "What other reforms," Taylor asked, "could do as much toward promoting prosperity, toward the diminution of poverty, and the alleviation of suffering?"[34] Taylor thought that increasing efficiency would be more important than such traditional reform staples as the tariff, the control of large corporations, and "more or less socialistic proposals for taxation."[35] But the supposed absolutism of Taylor's system probably enhanced the authoritarian tendencies already present in scientific management. Taylor insisted that individuals must live according to the laws of management, whether this accorded with their wishes or not. Scientific management was not something to be debated or voted upon; it was, in Taylor's view, a fiat of nature.[36]

Taylor's followers were convinced that they had the key to social reconstruction. As a result, their efforts to spread the new gospel took on some of the aspects of a religious crusade. Scientific management burst on the national scene in this guise in 1910 when Louis D. Brandeis gave national publicity to the movement in hearings before the Interstate Commerce Commission. One of the commissioners commented to a member of the group that "this has become a sort of substitute for religion with you." Ray Stannard Baker, noting the Taylorites' "extraordinary fervor and enthusiasm," wrote that "theirs was the firm faith of apostles."[37] Morris L. Cooke held that the goal of scientific management was to enable each individual to realize his highest potentialities; a society so organized, he thought, "approximates the millenium."[38] Scudder Klyce considered scientific management to be "*the* practical religion."[39]

The transformation of scientific management into a national reform movement had been underway for several years. But the year 1910 marked a new departure. Brandeis's publicity and the publication in the same year of *The Principles of Scientific Management* served to give a new direction to the campaign for efficiency. In both cases Taylor and his followers shifted their appeal from the business community to the general public. Taylor now called for the public to impose scientific management on business and the unions in the name of justice.[40] It was this reorientation that, in part, caused the scientific claims of the movement to receive greater stress. Taylor had intended his system to be scientific from the start, but the need to appeal to a wider audience required a shift in emphasis. To some extent the Taylorites were attempting to capitalize on the changing climate of national opinion. The progressive ferment in national politics seemed to favor social reform. But the metamorphosis of scientific management also reflected the reaction of the Taylorites to the resistance that their system encountered on the part of business. In their new mood, the devotees of scientific management attempted to spread it not only to industry, but to education and government as well.

Something of the fervor of Taylor's followers was indicated by their efforts to apply scientific mangement to education. In 1910, Morris L. Cooke did a study of efficiency in American higher education for the Carnegie Corporation. It created a furor. Leading educators protested in scathing terms against the alleged inaccuracies and distortions of Cooke's report. Cooke stood his ground, and on many matters of factual detail he proved to be better informed than the college administrators. But the real issue was the underlying spirit of the report. Cooke's procedure was simply to compare the workings of American universities with an ideal model of scientific-management administration. In this light, such fundamentals as academic freedom and tenure appeared to Cooke to be no more than barriers to efficiency. Cooke concluded that the university professor was a sort of "grafter" who, like the crooked politician, saw the aim of the whole system in his own economic security. Cooke was convinced that a revolution was required and that it would have to come from the outside. The aim would be to restructure these institutions in the model of an efficient factory. In a letter to the Carnegie Corporation, Cooke brushed aside complaints with the comment that his detractors did not understand that "for the art of management there has been developed a body of scientific laws as trustworthy and as much to be reckoned with as are those of engineering."[41]

Cooke's mixture of Messianic reform with intolerance to traditional educational practices was more than matched by Frank Gilbreth. When Gilbreth was asked in February, 1919, to speak during a meeting of the Society for the Promotion of Engineering Education, he did not attempt to conceal the contempt he felt for the assembled educators. "I think that you are a sorry lot," he exploded, "I think you do not dare to face the facts; I think this is a mutual admiration society, where you come together with the idea of stealing half of the wrong ideas from each other." He threatened that unless they forthwith sought to discover the "One Best Way," he would do it himself.[42]

Politics presented an even greater challenge than either education or industry. The expansion of scientific management to government raised a fundamental issue implicit in engineers' discussions of a larger social role for members of their profession. How could engineers exercise the necessary authority within a democratic society? Here again the great merit of scientific management was not that it solved this problem, but that its practitioners were forced to face the issue squarely.

The followers of scientific management developed not one but two theories of how to combine democracy and efficiency. One was simply to abolish democracy. This was done by redefining the term "democracy" to make it synonymous with "scientific management." This was the position of Henry L. Gantt. Morris L. Cooke took a different view. He attempted to reconcile democracy and efficiency by administrative innovations. Neither of these formulations had reached maturity before Taylor's death in 1915. Gantt's could be considered the more orthodox, since it commanded the larger support among the Taylorites. On the other hand, before his death, Taylor had indicated privately a strong preference for Cooke's interpretation of the philosophy of scientific management.[43]

Gantt was one of Taylor's ablest followers. He made two major contributions to scientific management. His method of wage payment proved to be more workable than Taylor's own scheme. Gantt's second innovation was a system of charts for measuring the daily performance of men and machines. By extension, the same graphical methods could be applied to whole plants and entire industries. They encouraged Gantt to consider the larger social implications of scientific management.[44]

Gantt was deeply concerned with national efficiency. He thought that the community should not have to bear the costs of the incompetence and wastefulness of the "financiers" who controlled American industry. After an ASME meeting in December, 1916, at which Gantt and one of his associates presented some of their views, an informal gathering of fifty mechanical engineers was held. They founded an organization, the "New Machine." Its aims were declared to be "the acquirement of political as well as economic power." Almost its only official act was to send a letter to President Wilson in February, 1917. In this letter Gantt and his followers argued that America's industrial system would grow "only through a progressive elimination of plutocracy and all other forms of arbitrary power." The New Machine called for a political union of all productive persons in order to free industry from the burdens of unearned income. They wished to "take the control of the huge and delicate apparatus of industry out of the hands of idlers and wastrels and to deliver it over to those who understand its operations." They thought that "the tools must belong to those who know how to use them."[45]

The New Machine was short-lived because its members soon were involved in war work. Gantt's own experience served further to convince him of the feasibility of planning for industry. During the war he introduced his charts for measuring the progress of arms production and ship construction. Gantt wanted to apply scientific management to the entire economy. To do this he called for the continuation and expansion of planning after the war. Industry would be organized into public service corporations and run for production rather than profits. Gantt's own charts would provide the basic tools for central planning. Power would be vested according to knowledge and ability, which, for Gantt, meant that engineers would hold most of the positions of authority. What Gantt proposed, therefore, was a technocracy. [46]

Gantt's proposed reorganization of society was incompatible with both liberty and democracy. "I don't agree with the average politician's concept of democracy," Gantt asserted. "His is the debating-society theory of Government." Gantt objected that under such a system policies were determined "not according to the laws of physics but by majority vote." To Gantt "real democracy consists of the organization of human affairs in harmony with natural laws." [47] Gantt also rejected liberty. "The new democracy," he explained, "does not consist in the privilege of doing as one pleases, whether it is right or wrong, but in each man's doing his part in the best way that can be devised from scientific knowledge and experience." [48]

It is important to note that Gantt did not propose to jettison democracy and liberty in the name of efficiency, or even science. His aim, like that of Taylor, was moral. "Efficiency alone," Gantt maintained, "will not cure our troubles." Efficiency was a means which might be "beneficial or detrimental as the end is worthy or unworthy." [49] The end Gantt had in view was to establish justice and end conflict between classes. American capitalism under the control of plutocrats had ceased to render service; it had become an engine of social injustice. Gantt thought that the application of scientific management would restore justice. Under it, each person would get precisely what he deserved, since scientific management provided an exact measurement of justice. To Gantt this was conservative. It was the one sure way to prevent a revolution and preserve the existing industrial system. It would keep the "extreme radicals" out of power. It would preserve not only existing institutions like private property but also traditional beliefs. Gantt hoped that through scientific management Christianity would be changed from a "weekly intellectual diversion to a daily practical reality." [50]

In contrast to Gantt, Morris L. Cooke saw in scientific management the fulfillment rather than the negation of democracy. Perhaps the difference was that Gantt was politically naive, whereas Cooke was the most politically literate of Taylor's followers. He had been active in Philadelphia reform politics for more than a decade before meeting Taylor. Appointed Director of Public Works for the city of Philadelphia, Cooke struggled to bring scientific management into meaningful relationship with American democracy. But however much he differed with

Gantt on the means of applying scientific management to government, he agreed with him that the ultimate significance of scientific management was moral. "We shall never fully realize either the visions of Christianity or the dreams of democracy," Cooke maintained, "until the principles of scientific management have permeated every nook and cranny of the working world."[51]

The task of reconciling scientific management and democracy was not easy. On the one hand, Cooke insisted that democracy and universal suffrage were "two undebatable questions."[52] He thought that democracy should prevail not only in government, but in industry and the engineering profession as well. In all cases, Cooke held that democracy must consist of "not only the rule of the people, but the participation by the people in their government in the largest possible fashion."[53] But Cooke also shared the elitism inherent in scientific management. In orthodox Taylorite fashion, he divided all questions into those involving "fact" and those that were matters of "opinion." The former should be determined by experts; only the latter should be voted upon.[54] He advocated "an absolute reliance on experts in every part of the municipal field."[55] Of all specialists it was the engineers who were destined to play a leading role in the politics of the future. Cooke foresaw a dominant participation by engineers in government, a "preeminence by consent" that would be "more than an equivalent for sovereignty."[56]

Cooke thought he saw a means of harmonizing these conflicting demands. The solution lay in administrative mechanisms that combined majority will with a larger degree of independence for the elected official. He saw a trend toward this sort of "administrative individualism" in such devices as the city manager, the short ballot, longer terms for administrative officials, smaller legislative bodies, and the use of judicial decision rather than juries for certain types of litigation.[57] Thus, Cooke imagined that the engineer would rise to power by popular election within the existing political framework. Once in office, he would be granted sufficient power to apply his expert knowledge without interference.

Cooke was aware that increasing administrative authority created dangers. Experts might abuse their power. For this the initiative, referendum, and recall provided protection for the public interest. More difficult was the question of public participation. Cooke believed in government by the people as well as for them. He argued, however, that there were democratic safeguards within scientific management, since consent and participation were inherent in its practice. Cooke thought that engineers must be dedicated to the public welfare. They should use publicity to inform the public, engage their interest, and win their active cooperation.[58]

Cooke realized that these measures would not be entirely adequate to protect democracy. Only engineers could fully evaluate the work of engineers. Here he saw a new and important function for the professional society, as guardians of the public interest. He also wanted the national engineering societies to encourage the larger application of engineering to society. On both counts it was particularly necessary that these associations be democratically governed and dedicated to the public interest. But, as he soon discovered, engineering societies were undemocratic and subservient to private interests. Like other engineering progressives, Cooke found that before he could reform society, he had first to reorganize his own profession.

NOTES

1. The standard biography of Taylor is Frank B. Copley, *Frederick W. Taylor, Father of Scientific Management*, 2 vols. (New York, 1923). For Taylor's own recollections of his early career, see Taylor to M. L. Cooke, December 2, 1910, file "Cooke, June–December, 1910," Frederick William Taylor Collection, Stevens Institute of Technology, Hoboken, New Jersey.

2. On the importance of apprenticeship training for elite mechanical engineers, see Monte A. Calvert, *The Mechanical Engineer in America, 1830–1910* (Baltimore, 1967), 3–23, *passim*.

3. Joseph W. Roe, *English and American Tool Builders* (New Haven, 1916), 250, 277.

4. Raymond Harland Merritt, "Engineering and American Culture, 1850–1875" (Ph.D. thesis, University of Minnesota, 1968), 94–131, surveys some analogous efforts by civil engineers to make railroad management more scientific.

5. Frederick W. Taylor, *Shop Management* (New York, 1919), 35. See also Frederick W. Taylor, *The Principles of Scientific Management* (New York, 1911), 13–14.

6. Taylor, *Principles*, 49–58.

7. For the origins and spread of the Taylor system, the best sources are Taylor, *Principles*, and Copley, *Taylor*. For two recent studies see Milton J. Nadworny, *Scientific Management and the Unions, 1900–1932* (Cambridge, 1955), 1–20, and Samuel Haber, *Efficiency and Uplift, Scientific Management in the Progressive Era, 1890–1920* (Chicago and London, 1964).

8. Taylor, *Shop Management*, 38–46.

9. *Ibid.*, 148–181. For a succinct summary see F. W. Taylor, "The Present State of the Art of Industrial Management, Discussion," *Trans ASME*, XXXIV (1912), 1198–1200.

10. Taylor, *Shop Management*, 110.

11. *Ibid.*, 92–98.

12. *Ibid.*, 99, 102.

13. *Ibid.*, 108–109, 126–127.

14. Taylor to L. P. Alford, August 21, 1912, file "ASME . . . 1912," and M. L. Cooke to Taylor, October 7, 1907, file "Cooke, 1907," Taylor Papers.

15. "Taylor's Testimony before the Special House Committee," in Frederick W. Taylor, *Scientific Management* (New York and London, 1947), 31.

16. The term "scientific management" was not officially adopted until 1910, and Taylor put greater stress on the "scientific" character of his system after 1910 than before. But he apparently assumed that his work was scientific from the start (see, for example, Taylor, *Shop Management*, 21, 45, 58, 60–63, *passim*).

17. Taylor to Alford, August 15, 1912, file "ASME . . . 1912," Taylor Papers.

18. "The Present State of the Art of Industrial Management," *Trans ASME*, XXXIV (1912), 1140.

19. J. M. Dodge to Taylor, January 18, 1915, and Taylor to Dodge, January 27, 1915, file "Dodge, 1891–1915," Taylor Papers.

20. The first modern scholar to stress this point was Hugh G. J. Aitken, *Taylorism at Watertown Arsenal* (Cambridge, 1960), 20–21. For an example of Taylor's ideas on this subject, see Taylor, *Principles*, 142–143.

21. Taylor, *Shop Management*, 25.

22. For a recent commentary on this belief, see Rene Dubos, *The Dreams of Reason, Science and Utopias* (New York and London, 1961), 6–12, *passim*.

23. Henri Poincaré, *Mathematics and Science: Last Essays* (New York, 1963), 103.

24. Klyce to Taylor, June 6, 1911, file "Scudder Klyce," Taylor Papers.

25. "Scientific Management and the Moral Law" (typescript), file "Scudder Klyce," Taylor Papers.

26. Taylor to M. L. Cooke, June 27, 1911, and Taylor to Ray Stannard Baker, June 26, 1911, file "Scudder Klyce," Taylor Papers.

27. Taylor to Klyce, June 26, 1911, file "Scudder Klyce," Taylor Papers.

28. Klyce to Taylor, June 6, 1911, file "Scudder Klyce," Taylor Papers.

29. E. D. H. Klyce, "Scientific Management and the Moral Law," *The Outlook*, IC (November 18, 1911), 659–663. This article deserves to be considered a joint product of Klyce, his wife, and Taylor; Taylor not only helped to edit it, but he added some illustrative material of his own (Taylor to Ernest H. Abbott, September 29, 1911, file "Scudder Klyce," Taylor Papers).

30. Edna Yost, *Frank and Lillian Gilbreth* (New Brunswick, 1949), 261–263. Gilbreth's work was not generally accepted, and most of the scientific-management group preferred to stick to Taylor's original methods (Milton J. Nadworny, "Frederick Taylor and Frank Gilbreth: Competition in Scientific Management," *Business History Review*, XXXI [Spring, 1957], 23–24).

31. Taylor, *Principles*, 8.

32. Haber, *Efficiency and Uplift*, 41, and Yost, *Gilbreth*, 223, 260. Frank Gilbreth is most closely associated with the "One Best Way." But the idea, if not the term, was implicit in Taylor's notion of scientific law, and it was used by Morris L. Cooke and other members of the scientific-management group.

33. Taylor, *Principles*, 5–8.

34. *Ibid.*, 14.

35. *Ibid.*

36. Taylor left it for his followers to work out the full social implications of his philosophy, but the general trend was clear enough in Taylor's own writings. For examples, see Taylor, *Shop Management*, 25, and Taylor to C. Bertrand Thompson, December 30, 1914, file "Colleges and Universities," Taylor Papers. The antidemocratic aspects of Taylorism have been stressed in Haber, *Efficiency and Uplift*.

37. Ray Stannard Baker, "Frederick W. Taylor—Scientist in Business Management," *American Magazine*, LXXI (March, 1911), 565.

38. Morris L. Cooke, "The Spirit and Social Significance of Scientific Management," *Journal of Political Economy*, XXI (June, 1913), 485–486.

39. Klyce to Taylor, June 6, 1911, file "Scudder Klyce," Taylor Papers.

40. Taylor, *Principles*, 138–139, 144.

41. Cooke to Dr. Henry S. Pritchett, February 12, 1910, file "Cooke, 1910"; see also Cooke to Taylor, April 3, 1914, file "Cooke, January–June, 1914," Taylor Papers, and Morris L. Cooke, *Academic and Industrial Efficiency* (New York, 1910), 22–23, 30–31.

42. Frank B. Gilbreth, "The Training Required for Engineers, Discussion," *Engineering Education*, IX (February, 1919), 238–239.

43. Taylor to Cooke, May 2, 1910, file "Cooke, January–May, 1910," and Taylor to Cooke, March 29, 1913, file "Cooke, 1913," Taylor Papers.

44. Leon P. Alford, *Henry Laurence Gantt, Leader in Industry* (New York, 1934), 85–94, 207–223, and Wallace Clark, *The Gantt Chart* (New York, 1954).

45. Alford, *Gantt*, 264, 270–273. There is a copy of the New Machine's letter to the President in file 83, box 58, Papers of Morris L. Cooke, Franklin D. Roosevelt Library, Hyde Park, New York.

46. Henry L. Gantt, *Organizing for Work* (New York, 1919), 8–10, 18–22, 58–64, 72–73, 91, 106.

47. New York *World*, June 30, 1918 (editorial section), p. 1 (quoted in Alford, *Gantt*, 259–260).

48. Quoted in Alford, *Gantt*, 196.

49. H. L. Gantt, "The Relation Between Production and Costs, Discussion," *Journal ASME*, XXXVII (August, 1915), 475.

50. Gantt, *Organizing for Work*, 109; see also 3–8, 11–15, 101–108.

51. Morris L. Cooke, "The Spirit and Social Significance of Scientific Management," *The Journal of Political Economy*, XXI (June, 1913), 493.

52. Morris L. Cooke, *Impressions of an Engineer in Public Service* (Philadelphia, 1915), 6.

53. Morris L. Cooke, "The Spirit and Social Significance of Scientific Management," *The Journal of Political Economy*, XXI (June, 1913), 490.

54. Morris L. Cooke, "Some Factors in Municipal Engineering," *Journal ASME*, XXXVII (February, 1915), 83.

55. Cooke, *Impressions of an Engineer in Public Service*, 10.

56. Morris L. Cooke, "On the Organization of An Engineering Society," *Mechanical Engineering*, XLIII (May, 1921), 323.

57. Morris L. Cooke, "Who is Boss in Your Shop?" *The Annals of the American Academy of Political and Social Science*, LXXI (May, 1917), 170.

58. *Ibid.*, 170, 181. See also Morris L. Cooke, "Public Engineering and Human Progress," *Journal of the Cleveland Engineering Society*, IX (January, 1917), 245–263.

7 ▪ MORRIS L. COOKE:

THE ENGINEER AS REFORMER

Scientific management had a profound influence on the entire engineering profession. Within the ASME, Taylor's followers constituted a ready-made reform faction. As a result of their activities, the ASME was the most thoroughly revolutionized of American engineering societies. By 1919, the Taylorites momentarily seemed to control the society. They caused the ASME to take the lead in persuading the founder societies to replace the Engineering Council by a more representative body, the Federated American Engineering Societies. When this new organization attempted to bring engineering to bear on the great national questions of waste and labor unrest, it was the scientific-management engineers who provided the expert knowledge and leadership. Although few engineers outside of the ASME were converted to scientific management, it nevertheless provided inspiration and guidance for other progressives. The failure of scientific management to finally achieve its ambitious professional and political ends did much to bring about the ultimate collapse of engineering progressivism as a social movement.

The success of the Taylorites as reformers was a consequence of the transformation of mechanical engineering. Although few in numbers, the scientific-management group could exploit a growing rift in the society. Representatives of the older order, who were apt to be found in the machine-tool and light-manufacturing industries, were committed to the idea that mechanical engineers should be independent, self-employed practitioners. The new generation of organization men identified the engineer's destiny with the large corporations in which engineers served as employees. Taylor himself was a member of the older elite, and he could draw upon a network of friends outside his own efficiency group. But more important was the fact that in the ASME context, scientific management provided a powerful ideology that promised to restore independent practice and defend professional values. Both appeared to be threatened by commercialism.[1]

In its first two decades the ASME had been controlled by an informal oligarchy of creative professionals. But by 1904, the older elite had lost its dominant position. In that year the society devised a more formal, bureaucratized governmental structure, the most conspicuous feature of which was a shift in power from the council to a series of standing committees. These powerful committees tended to be staffed by men whose distinction arose more from bureaucratic rank than from important technical innovations. The creative were few in number and geographically scattered. As a practical matter, staffing the committees required a comparatively large number of engineers residing close to the society's New York headquarters. This tended to enhance the influence of men associated with the great corporations located in the New York area.[2]

Taylorism was fundamentally at variance with the new order of things in the ASME as well as in mechanical engineering generally. Taylor did not accept the bureaucratized, commercialized nature of much contemporary practice; his system was a means of restoring professionalism and independent practice. His system was, by implication, no less at odds with the ASME's new government. Scientific management rejected the committee system of management adopted by the ASME, and it favored the centralization of power. Quite apart from his doctrines, Taylor, as a member of the old elite, disliked the new regime. In private he expressed his contempt for the financiers and their representatives in the ASME. He thought that if the society's new rulers treated its older leaders just like other members, "they then range the society right among the hogs."[3] Two events helped to make manifest the latent antagonism between scientific management and commercialized engineering. The first was Taylor's presidency of the ASME in 1906. The second was the efficiency group's criticisms of railroad management in their testimony before the Interstate Commerce Commission in 1910.

When Taylor became president of the ASME, he attempted to apply scientific management to the society itself. Since 1902, his followers had been arguing that the society's financial difficulties could be solved by greater efficiency. No doubt they hoped to make the ASME a showcase of scientific management. Taylor concentrated on the society's administrative machinery rather than on its governmental structure. However, a measure of administrative centralization was achieved by replacing the part-time secretary, Hutton, by a full-time man, Calvin Rice. Taylor was able to make important savings in publication costs and improvements in office routine. But his reforms aroused a storm of protest. Hutton,who had been forced to resign, and the permanent staff of the society resisted bitterly. They had powerful friends among the membership. Opponents of Taylor complained that costs had gone up rather than down. The value of Taylor's administrative innovations remained a source of controversy within the society for many years.[4]

The result of Taylor's presidency was probably to make acceptance of scientific management by the society more difficult. Although Taylor and others of his followers served on the ASME's governing council, they met strong resistance when they tried to influence the society's policies along lines suggested by scientific management. In January, 1909, President Jesse M. Smith "read the riot act" to Taylor, serving notice that he intended to oppose any proposal made by him. All suggestions for change, Smith insisted, must come from the committees or the membership at large. Rather than start an open war, Taylor absented himself from meetings of the council for a year.[5]

The hearings before the Interstate Commerce Commission in 1910 were a turning point for scientific management. They not only gave national publicity to the movement, but they tended to galvanize opposition to scientific management within the ASME. Henry L. Gantt, who was serving on the governing council, attempted to get the ASME to endorse Brandeis's contention that increased efficiency would make a rate increase for the railroads unnecessary.[6] Not only was he defeated, but the ASME came close to repudiating scientific management between 1910 and 1912. In 1910, the ASME refused to publish Taylor's *Principles of Scientific Management*. In 1912, a committee rejected the scientific claims of the Taylor system. The mounting resistance to the publication of any papers on scientific management was one reason for the formation by Taylor's followers of a technical society of their own.[7]

The man who succeeded where Taylor, Gantt and others failed was Morris L. Cooke. He had been hired by Taylor to assist in systematizing the ASME's publication policies. During his year working with Taylor on the ASME's administrative machinery, he had an unrivaled view of the inner workings of the society. His knowledge of the location of the hidden levers of power was to prove invaluable. For it was Cooke who, between 1914 and 1919, carried through a series of sweeping reforms that shook the society to its foundations and that spread beyond the ASME to the entire engineering profession. The reason for his success was twofold. He had a gift for politics and intrigue. He grasped the key fact that the hard core of resistance to scientific management came from the public utilities and railroads acting together as a sort of monopoly interest within the engineering profession. By shifting emphasis from the virtues of scientific management to the vices of the utilities, he was able to broaden the base of his appeal and link the efficiency crusade to the national progressive movement.

The position of Morris L. Cooke in the scientific management movement was anomalous. Although he was one of the last to be admitted to Taylor's inner circle, he was Taylor's favorite. Yet Cooke was only marginally an engineer. He had received a good engineering training, both at Lehigh and as an apprentice and journeyman machinist. But he had practiced his profession for only a short time before going into business. From 1899 to 1905, he had been associated with the printing and publishing trades in various capacities. Not until 1905 did he return to engineering as a disciple of Taylor's. He lacked the quality of inventive genius so conspicuous in Taylor, Barth, Gantt, and Gilbreth. His technical achievements were modest.[8]

Cooke more than made up for his technical limitations. A reformer by temperament and upbringing, Cooke's interest lay in the broader social and political implications of scientific management. For this Cooke was well prepared. He had worked as a newspaperman, reformer, and publisher. He was able to give good editorial advice to Taylor. His vast acquaintance in journalistic and reform circles assisted his publicity work. His sensitivity to the world outside the factory enabled him to see deeper implications of scientific management. Taylor heartily approved of Cooke's optimistic, reformist gloss of his doctrine.[9] Cooke was one of those responsible for the adoption of the term "scientific management" to describe the movement.[10] He encouraged Taylor to stress the universality of the laws of management.[11] Finally, it was Cooke who attempted to apply scientific management to government when serving as Director of Public Works for the city of Philadelphia.

Although Taylor prized Cooke for his unique abilities, another reason for their intimacy was similarity in background. Both came from the same upper stratum of the same city, both had attended exclusive private schools, and both had departed from family tradition to enter engineering. Cooke's father had been a physician, and his immediate relatives were members of the learned professions. But despite its culture and distinguished lineage, the family was overshadowed by the newly rich industrialists who emerged in the gilded age. Cooke's father had been a mugwump reformer, and he indoctrinated his children with ideals of public service and political activism. A crisis in family finances that forced Cooke to leave college and work for a time may have reinforced these interests. As a young man Cooke joined the sporadic, if rather futile, efforts of the good government men to restore honesty to the politics of their native city.[12]

To some extent Cooke may have seen scientific management as a new way of achieving the goals of mugwump reform. It was a means of transferring political leadership from corrupt bosses to educated gentlemen, suitably insulated from popular majorities. It promised cheaper and more efficient government. But Cooke differed from the older generation of reformers in important respects. Where the mugwumps had faith in a class, Cooke looked to his profession for social salvation. Cooke's belief in the absolutism, universality, and determinism of the scientific laws of management gave his thought a Utopian cast missing from the genteel doctrines of the mugwumps. Although Cooke wished to limit the number of issues decided by the people directly, his appeal was for the voters to rally against the misdeeds and inefficiency of selfish business interests. This basic strategy set Cooke apart from not only mugwump reformers, but from his elders in scientific management as well.[13]

Cooke's hopes for scientific management were closely bound to the ASME. He wanted the society to become a forum for scientific management and reform. He envisaged its role as "akin to the University of Wisconsin."[14] For this it was necessary that the ASME and its members be dedicated to public service. Even while still a junior member and ineligible to vote, Cooke sought to use his unrivaled knowledge of the ASME's inner workings. His first effort came in 1907. He organized a successful campaign to reelect Fred J. Miller to the society's council.[15] By 1908, he embarked on a grandiose plan to reorganize the ASME. In a paper, "The Engineer and the People," he laid down, in general terms, the basic philosophy that he was to expand and develop for the next twenty years and more. He saw public service as the key to the engineer's status. The low status of the profession was a consequence of the fact that it did not serve the public directly. But engineers might act through their profession to accomplish collectively what they could not do individually. Cooke proposed that the ASME establish a committee through which engineers could serve the public.[16]

Cooke's paper was intended as the first step in a carefully calculated campaign. As he was to do later, he prepared the way with great care. He lined up supporters in advance to make favorable comments, and he drafted a resolution to be introduced by one of his supporters which called for action on his proposals by the ASME.[17] Rather to his surprise, Cooke's paper aroused no controversy. Leading conservatives were lavish in their praise.[18] Cooke's resolution passed without opposition, though it took until 1910 to amend the society's constitution. The reason for the unanimity was that Cooke's paper was rather general. The idea that the engineer's status could be raised by public service was common to both the conservative and progressive versions of professionalism.

But underlying Cooke's philosophy was a radical departure from older views of professionalism. What made Cooke's proposal revolutionary was his assumption that loyalty to the public and loyalty to employers were antagonistic, even incompatible. This theme was not stressed in Cooke's 1908 paper; it became clear only later. Cooke saw engineering and business as radically different. He resented "the assumption that business—big or little—is engineering."[19] His animosity was directed particularly toward the large, monopolistic corporations, which he felt were exploiting the people and polluting American politics. Engineers who allied themselves with such organizations were acting against the public interest. "Engineering has a great future," Cooke held, "but engineering dominated by trade is a terrible menace to society."[20] Cooke was convinced that it was the alliance of engineering with predatory wealth that was responsible for the low status of the profession. To Cooke, therefore, the remedy lay in public service. As he wrote to a friend:

> The more I think of it, the more I feel that the fundamental consideration in the work of an engineer—if he is ever to pull himself out of his present status of being a hired servant—is that he shall make public interest the master test of his work.[21]

Even before presenting his paper on the theme of social responsibility, Cooke was busy organizing the second step in his campaign. This was to be a meeting of the ASME on smoke abatement. His proposal was a radical departure from the usual practice. Its purpose was more reform than the advancement of knowledge. The papers were to be as nontechnical as possible. The entire session was to be highly structured, being aimed at describing the existing evil and pointing to a remedy.[22] As before, Cooke laid careful plans, sending out letters, drafting a circular, and lining up support. Despite these preparations, Cooke's proposal was turned down by the Meetings Committee. Among the largest offenders in the matter of air pollution were the various large utility companies. They were well represented on the committee. Although corporate interests helped to defeat Cooke, the committee members had a point when they argued that Cooke's proposed meeting was not within the scope of the ASME.[23]

Cooke's first attempt to reform the ASME, in 1908 and 1909, was premature in more ways than one. Not only was the society not yet ready for such a radical departure, but neither was Cooke. As a junior member without technical reputation, he was in a very poor position to influence his profession. Indeed Cooke's absorption with politics and other matters almost wrecked his career as a management consultant. He had secured a contract to systematize a printing company, and Taylor had agreed with the owners to underwrite any losses that they might suffer by retaining a fledgling.[24] Cooke's attempt to apply scientific management was not successful in this case, and Taylor had to pay a rather substantial sum to the owners. Taylor blamed the failure on Cooke's spreading himself too thin.[25] Cooke took the criticisms in good part; and from 1910 to 1912, he devoted himself assiduously to scientific management. He passed up, for example, the opportunity to testify for Brandeis before the Interstate Commerce Commission.

Cooke's initial failure did have one lasting effect: it led to a rift with Henry L. Gantt that was never completely healed. Cooke brought Gantt in on his first job as a consultant. Gantt thought, apparently not without reason, that Cooke was making a mess of things; he also resented the fact that he, the senior man, should be taking orders from a novice. The attendant friction was magnified when Gantt proceeded to tell Cooke's employers just what he thought was wrong with the way Cooke was doing things. Taylor chided Gantt for this unethical conduct; it was Cooke who had the contract, and Gantt had been retained by Cooke, not by the owners.[26] In later years the relations between the two men were correct, but there was no real cordiality. Cooke and Gantt, therefore, attempted to develop the political and social implications of scientific management independently of one another.

Cooke got his opportunity to apply scientific management to government when he was appointed Director of Public Works for the city of Philadelphia in the reform administration of Mayor Rudolph Blankenberg. What emerged was something quite unlike the Taylor system applied to a factory. Cooke was not able to install an incentive wage, nor to engage in time studies. Without these there could be no precisely determined daily task for each workman. A planning board was created; but without time study or task setting, its functions bore little relation to the planning department of the Taylor system. Cooke used many of the separate mechanisms of scientific management, such as standardization, functional management, routing, and cost accounting. But the result was not the Taylor system. Despite some improvements in technical matters like snow removal, water purification, and coal handling, the most important changes did not involve the work of employees at all. Rather the primary emphasis in Cooke's administration lay in developing new relationships with the public and the private businesses serving the city.[27]

One of the most important things that Cooke tried to do in Philadelphia was to demonstrate the possibility of expert leadership in a democracy. It would be possible for a democratic government to match the efficiency of an autocracy if it had strong leadership. But, as Cooke pointed out, leaders needed followers. The leader had to win the support of the people. It was necessary for the people to acquire a "novel attribute," that of "learning how to support efficient leadership in more substantial fashion."[28] The solution, Cooke thought, lay in publicity. Indeed, Cooke rated his publicity work as the "most notable improvement" of his administration.[29] At his instigation, the city embarked on one of the most extensive publicity campaigns in American municipal annals. Cooke used every possible media to get across his message: posters, pamphlets, newspapers, comic strips, motion pictures, and exhibits. In many ways Cooke was successful; the voters were probably better informed than ever before. But that did not stop them from restoring the old machine at the end of 1915. One reason, possibly, was that Cooke's publicity was a one-way street. There were no effective means whereby the people's own wishes could influence policy. Cooke's mechanisms for the feedback of public opinion remained on the suggestion-box level.[30]

Cooke claimed to have saved the city more than one million annually during his four years in office. But the bulk of the savings did not come from scientific management; they were of a more traditional sort. The greatest source of inefficiency, as of municipal corruption, came from the systematic plundering of the city by an alliance of venal politicians and selfish business interests. Cooke, therefore, undertook a classical reform campaign to eliminate gross corruption and fraud. He uncovered evidence, which he published in detail, exposing the system of pay-offs by city employees to political bosses. Almost all city contracts had been made through a few powerful politicians and their friends. Cooke ended this practice. By means of open bidding, standardized specifications, and strict inspection procedures, Cooke was able to ensure good service at much lower cost. The largest single saving Cooke made, one-quarter of the total, was the result of a new contract that cut the price of the city's garbage collection in half.[31]

Cooke found that the powers of his office were only sufficient to deal with the smaller offenders. The major utilities presented a far more difficult problem. Their monopoly position made them insensitive to the sort of contractual checks that worked for competitive business. Their political power was very great; it extended beyond the city to the state and the nation. Yet the need to bring them under control was urgent. Cooke saw them as the principal source of municipal corruption. The excessive rates charged for electricity drastically limited use. Thus, not only was the city overcharged for street lighting, but the public was denied the fruits of technological progress. Cooke had a special reason for concern. It was the alliance of engineers with these predatory business interests that was responsible for the low status of his profession. Cooke had already had an inkling of their power in the ASME when engineers affiliated with the utilities torpedoed his proposed smoke-abatement meeting. The greatest battle of Cooke's term of office, and the turning point in his career, was, therefore, with the Philadelphia Electric Company.[32]

Cooke became convinced that the city was paying too much for its electric street lighting. He held several conferences with the company, but they were only willing to offer token concessions to the city and none to general consumers. Unable to obtain satisfaction by negotiation, Cooke proceeded to court. But here he ran into legal difficulties. Pennsylvania had been slow to undertake control of public utilities. Not until 1913 was the state's first public-service commission created. This provided the necessary machinery through which Cooke might act. But any serious challenge to the existing rates must be based on a valuation of the company's assets, and the public-service commission had no funds of its own for this purpose. The commission did have the authority to compel the company to conduct such a study, but only if the city first established a prima-facie case, which placed the burden on the city to conduct an expensive and lengthy preliminary investigation. The difficulty with this procedure was that it gave the company many opportunities for delay, misrepresentation, and sabotage.[33]

The Philadelphia City Council refused to supply the funds necessary to conduct a preliminary investigation. The council was in the hands of the old machine, and it was hostile to Blankenberg's reform administration. However, Cooke was not daunted. The act creating the public-service commission allowed a citizen to enter suit before the commission, and Cooke proceeded as an ordinary taxpayer. His investigation included two separate projects. An economist unraveled the tangled corporate history of the company and revealed an extensive watering of the company's securities. The second, and more difficult, task was that of finding an engineer to do a preliminary valuation of the company's assets. Here, Cooke encountered a problem; the consultants in this field were closely allied with the utilities and had absorbed their views and interests. They were not impartial, but partisans of the private interests against the public. After an extensive search, Cooke at last found a competent man in West Virginia. His valuation, though only preliminary, strongly suggested that the company's rates were excessive.[34]

Once Cooke had established his prima-facie case, the commission was in a position to order an inventory. The company, however, announced that an inventory was already underway under the direction of a nationally known engineer, Dugald C. Jackson, a dean of the Massachusetts Institute of Technology. Jackson's basic strategy was obfuscation. He deluged the commission with a vast quantity of undigested information. The inventory filled 110 volumes. Jackson included separately every pole used by the company; in one instance he listed separately a short length of two-inch pipe and the threads on either end. In contrast, the valuation itself was very short and in such a form as to make impossible a comparison between the inventory and the valuation. Had the commission accepted Jackson's work, it would have been all but impossible to get behind the data to the method of valuation. Fortunately for Cooke, the commission, recently reorganized by the governor, ordered that the valuation be submitted in the same form and the same detail as the summary of the inventory, so that the value placed on particular items could be discovered. When this was done, Cooke and his attorney had a field day. They were able to bring out many examples of the gross overvaluation of particular assets. Realizing that its case was hopeless, the company capitulated and reached a voluntary settlement. This involved drastic cuts in the price of electricity. The city saved $200,000 annually; the general public saved over $1,000,000 each year. Cooke's victory was complete.[35]

Cooke had won a battle, but not the war. The nationwide struggle between the cities and the utilities continued. The problem, Cooke thought, was that the cities were at a grave disadvantage. They were disunited, but the utilities were united. Cooke wrote to nearly three hundred mayors suggesting that the cities combine against their common foe. He got initial endorsement from approximately one hundred cities for a "national campaign" aimed at "curbing utility corporations."[36] Cooke then organized a conference of mayors, which was held at Philadelphia in November, 1914. The assembled dignitaries gave their blessings to a new organization, the Public Utilities Bureau, which

Cooke intended as the spearhead of the fight with the utilities. The bureau was to be supported by contributions from the cities. Its board of trustees included such prominent reformers as Louis D. Brandeis, Felix Frankfurter, Charles A. Van Hise, and S. S. Fels, as well as two engineers, Cooke and Taylor. In practice, Brandeis and Cooke were the moving spirits of the organization, although S. S. Fels made immediate action possible by underwriting the cost of the first year's operations.[37]

The Utilities Bureau was less successful than Cooke had hoped. Cooke had counted on the support of about two hundred cities, but only sixteen joined. Instead of an annual budget of $20,000, the bureau had to get along with only $5,000. Under the circumstances, the chief function of the bureau was to serve as a clearing house for information and a center of antiutilities publicity. It lacked the means to undertake anything more ambitious. Cooke, as manager, was able to put out a magazine that carried details of utilities cases across the nation.[38] At the end of the bureau's first year, Cooke organized a conference on valuation at Philadelphia. It constituted a sustained argument against the reproduction theory of value, cherished by the utilities since it enabled them to value old equipment at the same price as new replacements. Since the original cost theory of valuation favored by the conferees had not been accepted by the courts, the meeting had little practical effect.[39]

The disunity of the cities was only one aspect of the national utilities problem. The alliance of the engineering profession with the utilities was of even greater importance in Cooke's view. Concurrently with his organization of the Utilities Bureau, Cooke attempted to present the cities' case before the ASME. Cooke prepared a paper that laid out the problem. The cities were unable to obtain the services of impartial engineers, because those in the field were already committed to the utilities. Cooke pointed out that this alliance with selfish economic interests compromised the status of engineers. He thought that if the engineer was willing to make the public interest the guide to his practice, the community was ready "to accord the engineer a leading, perhaps controlling part" in government.[40] Cooke hoped to present his paper as part of a comprehensive review of the cities' engineering problems. He solicited a number of papers, most of which were relatively innocuous. But a paper by Frederick W. Ballard on Cleveland's municipal electric plant was another matter. Its conclusion, that a city-owned plant could sell electric power profitably at a fraction of the rate currently charged by private utilities, was highly controversial.[41]

Cooke faced a problem in attempting to present his views before the ASME. The publications committee was packed with engineers affiliated with the utilities, and they would never accept papers like his and Ballard's. Cooke resorted to a strategem that revealed his mastery of the details of society organization. He persuaded the dormant committee on public relations to sponsor his papers for a special meeting. He then was able to by-pass the publications committee; the meetings committee, which had a long-standing dispute with the publications committee, approved a special session on municipal engineering for the annual meeting in December, 1914.[42]

Cooke's appeal to his profession led to a bitter battle with representatives of the utilities interests. They protested Cooke's rather foxy tactics and attempted to prevent the papers from being presented. When this failed, they criticized Cooke's and Ballard's papers when they were presented. Ballard's figures on costs were severely attacked. But the harshest comments were reserved for Cooke. Alexander C. Humphreys, a prominent consultant in the field of gas engineering, led the assault. He held that anyone who would make such allegations of bias was "unworthy of a place in the profession."[43]

Cooke's meeting on municipal engineering turned into an election campaign within the ASME. Cooke had been nominated to serve on the governing council of the ASME. Humphreys and other council members affiliated with the utilities organized a campaign to defeat Cooke. Each side distributed to the membership circulars that obscured the issues. Both sides ignored the utilities in their campaign literature, relying on private discussion and correspondence to carry the message. The utilities based their case on a technicality in the manner of Cooke's nomination. They appealed to employers, urging them to canvass their engineering staffs. Cooke's official position was that his opponents did not favor engineers serving in public office. In his private correspondence, however, he made it clear that the issue was the domination of the ASME by the utilities. Aided by his own political virtuosity and the contacts of his colleagues and friends, Cooke was victorious.[44]

Cooke's experiences in the ASME and in Philadelphia stimulated him to undertake an investigation of the sources and extent of the power of the utilities in the engineering profession. Cooke identified three groups that served as carriers of utilities influence: employees and officers of utilities; engineers affiliated with their suppliers, such as the manufacturers of electrical equipment and steam boilers; and consultants whose practices depended upon the utilities. During his election campaign Cooke pointed out in private letters that the council which opposed his election included eight officers of utility corporations and three members whose businesses were dependent on the utilities.[45] The most vociferous spokesmen for the utility point of view were educators with private consulting practices. Alexander C. Humphreys, Cooke's chief antagonist in the ASME, was president of the Stevens Institute of Technology. But he was also the head of a consulting firm doing business with the gas utilities, and he was president of the Buffalo Gas Company and director of several others. D. C. Jackson, the engineer retained by the Philadelphia Electric Company in its fight with Cooke, was on the staff of the Massachusetts Institute of Technology. He also directed a large consulting firm that brought him several times the income he received as an educator.[46] Jackson and Humphreys were probably the most outspoken defenders of the utilities in the profession; both argued that it was the engineers' duty to defend public-service corporations.[47]

Although engineers affiliated with the utilities constituted only a small minority of the ASME's membership, they were able to exercise a disproportionate influence in its government. Cooke investigated the business affiliations of society officials, past and contemporary. The council that had opposed Cooke's nomination proved to be unexceptional. Cooke complained that "a very large percentage of the members of the Council, as it has been constituted during the last few years, have direct connections with public-utility corporations."[48] Cooke found a pattern of utilities representation on key committees also, particularly those concerned with publications and nominations. He was soon presented with an object lesson of how they operated. In 1915, the year following Cooke's election, all but one of the members of the new nominating committee were affiliated with the utilities. Their choice for president, duly ratified by the membership, was D. S. Jacobus. He was a professor at Stevens and an associate of Humphreys. Since 1906, he had been spending most of his time working for one of the largest steam-boiler manufacturers in the country, a major supplier of the utilities. Once in office, Jacobus proceeded to emasculate the Committee on Public Relations, by persuading all its members, save Cooke, to resign.[49]

The same pattern that Cooke found in the ASME applied, with minor variations, to all of the major engineering societies. In the case of the AIEE, the dominance of the utilities was notorious. But more surprising, however, was the case of the ASCE, the most professionally oriented of the founder societies. It had long been known that the railroads were able to exercise a veto in its affairs; Cooke found that by a sort of professional courtesy the utilities were granted the same powers. A special committee of the ASCE had been appointed to formulate principles and methods for the valuation of railroad and utilities properties. Although highly favorable to the railroads' point of view, the report was suppressed and the committee reorganized. In matters of common interest, the utilities acted together.[50]

Cooke thought that he had discovered a sinister conspiracy by the utilities to destroy the integrity of the entire engineering profession. At the center of this plot, Cooke saw the electrical utilities and their national representative, the National Electric Light Association.[51] Consultants in valuation work, like Jackson, were no longer independent, but hired mercenaries. He compiled lists of cases where their valuations were grotesquely inflated.[52] Independent consulting work in the construction of utilities was threatened by a corrupt alliance between the utilities and certain large engineering corporations. This raised the specter of the "ultimate unification of all engineering on a business rather than a professional basis."[53] The NELA was reaching out to corrupt engineering education, also, by paying professors fees for speaking engagements and offering them special associate memberships in the association.[54]

To Cooke the issue was clear. The NELA "now hangs like a millstone about the neck of American engineering, guiding its every step and thwarting almost every noble aspiration."[55] The engineering profession must choose between business and public service. The remedy would be to adopt sufficiently stringent ethical standards so that the utilities would be "put under absolute control," and the quacks allied with them driven from the profession.[56] If this were done, effective public regulation of the utilities would become possible. Only in this way could public ownership be avoided. The utilities would be directed away from financial manipulation to operating efficiency. The engineers, as champions of the public, would enjoy a vast increase in status and power.[57]

Cooke attempted to awaken his profession and the public in a series of public lectures, which he published as a pamphlet, *Snapping Cords*. In his critique, Cooke blended themes from traditional reform and scientific management. The utilities had grossly watered their securities to benefit financial manipulators. This led to high rates, which they defended by bribery of public officials. The utilities were also inefficient. The utilities' overlords were ignorant of costs and relied on crude, rule-of-thumb methods of management. Employees were tyrannized by a military-style bureaucracy. Cooke was particularly incensed at the complicity of engineers in the utilities' conspiracy. In his previous paper before the ASME he had criticized, in very general terms, the bias of consultants in the utilities field. Now he was painfully explicit. He attacked by name four of the leading educators whose private consulting work brought them into close alliance with the utilities interests. They included his old foes Humphreys and Jackson, as well as Mortimer E. Cooley and George F. Swain, affiliated with the University of Michigan and Harvard, respectively. In each case, Cooke presented detailed evidence of bias in favor of the utilities. For example, Jackson had been employed in 1909 by the city of Chicago to appraise the Chicago Telephone Company. He had reported that the company was running at an annual loss of over $800,000. The city had rejected Jackson's report, and after a fresh study, the company consented to a reduction in rates amounting to $700,000 yearly.[58]

Not content with assaulting these men's professional reputations, Cooke proceeded to add injury to insult. Cooke distributed copies of *Snapping Cords* to prospective employers of these engineers. The city of Passaic, New Jersey had, in fact, already retained Cooley to make an inventory of the street railways of that city, and Jackson was under retainer to make an inventory of the Public Services Corporation. Cooke wrote offering to supply information that would break any contract.[59]

The men attacked by Cooke demanded that he be punished; Cooke's actions violated the ASME's code of ethics. But he thought that the existing codes of engineering ethics had been "dictated" by "those who have debauched our cities and by those whose god is the dollar."[60] Cooke was probably courting an ethics suit, in order to demonstrate his contention that a new set of principles was needed to free the profession from commercial control. To Brandeis he wrote, "a 'heresy' trial in engineering is something new and I am anxious that the first one shall be carried along so as to make the ecclesiastics green with envy."[61] Cooke demanded a public trial and the right to cross-examine witnesses. With Louis D. Brandeis as his attorney, Cooke would be able to turn such a trial against his accusers; it would have been a splendid forum from which to present his case to the engineering profession. The officers of the ASME were not willing to oblige Cooke. They were not anxious for the sort of public confrontation that Cooke demanded. After Cooke's term on the council had expired, the council voted a mild censure of Cooke. Cooke threatened to take the society to court if the censure was not expunged from the record. The council meekly obeyed.[62]

Cooke, however, was not to be denied his appeal to the ASME. He published and distributed to the entire membership a second pamphlet, *How About It?* in which he denounced the "absentee management" of the ASME by business interests. The government of the society, Cooke charged, was "archaic, undemocratic, and inefficient." Its inner operations were shrouded in secrecy; published reports and minutes of meetings were fragmentary and misleading. Membership control was virtually impossible. Instead, real power, Cooke maintained, was vested with the New York "junta" representing the private utility interests. To substantiate his charges Cooke publicized many matters previously known to only a few. He reviewed, in detail, his own ethics case. He exposed the sabotaging of the Committee on Public Relations. He appealed to the membership to recast the society's government to make it democratic.[63]

Cooke's muckraking assault prepared the way for important changes. Although prominent leaders were outraged and humiliated by Cooke's attacks, they were unable to retaliate because Cooke received a great deal of support from rank-and-file members. Control by New Yorkers and the alliance of elite engineers with employers were long-standing complaints in all of the founder societies. Cooke's timing was excellent, because the discontent of ordinary engineers was increasing due to a combination of wartime inflation and rising expectations.

Ira N. Hollis, who succeeded Jacobus as president in 1917, tried to appease the dissidents. While denying that the society was undemocratic, he admitted that the directors were only human and that mistakes had been made.[64] Hollis and other ASME leaders made important concessions in response to Cooke's charges. Beginning in 1917, the selection of the nominating committee was informally turned over to the local sections; by 1920, it had been made elective. More younger men were appointed to standing committees, and efforts were made to secure a better geographical distribution in appointments.[65] Cooke's influence was not confined to the ASME. His protests at the throttling of the Committee on Public Relations produced results. President Hollis held that public affairs were properly the concern of the entire profession, and he approached the other founder societies to propose some form of joint action. Hollis's initiative coincided with mounting concern in the ASCE over the growth of the AAE and the restiveness among civil engineers generally. The result was the formation of the Engineering Council; Hollis became its first president.[66]

Cooke followed up his advantage. He pushed a campaign to eject the NELA from the tax-exempt Engineering Societies Building. The NELA represented everything Cooke most despised. Nominally a technical society, the NELA was in fact the national spokesman for the electrical utilities. Thus, it symbolized the alliance of engineers with predatory business, which Cooke saw as the root cause of all of the profession's difficulties. The NELA, of all utilities' associations, was the most involved in municipal corruption; Cooke called it the "strongest single arm" of "invisible government."[67] Cooke had begun his campaign as early as 1915 by private protests. He brought the issue into the open with his pamphlets *Snapping Cords* and *How About It?* Cooke gained very substantial support among engineers, not only in the ASME, but from the ASCE as well. And since a unanimous vote by all four of the founder societies was required, the NELA stayed where it was. But Cooke won a moral victory. Not only did he arouse the membership, but he denounced the light association to the New York City Tax Board. As a result, the city levied a tax on that part of the engineers' building occupied by the NELA. He had succeeded in branding it as a commercial organization.[68]

The climax of Cooke's campaign was intended to be the adoption of a new code of ethics that would divorce engineering from commercialism. By 1918, the time appeared ripe. The membership, in part due to Cooke's propaganda, was restless and demanded sweeping changes. Cooke, as usual, prepared his case with care. His paper, "The Public Interest as the Bed Rock of Professional Practice," was a sober, factual summary, in sharp contrast to most of his recent publications. In it, Cooke pointed out that none of the existing ethical codes of American engineering societies made the public interest paramount. Cooke was able to show that other professions not only put the public interest first, but that they did so in ringing words. Cooke did not need sensationalism; his parallel quotations from the codes of the engineers and those of doctors and lawyers constituted a devastating indictment. Possibly to avoid controversy, Cooke did not mention commercialism or the public utilities.[69]

Cooke got action, but not exactly the sort he had intended. Cooke's paper served to bring to the surface the seething discontent within the society. In the discussion, Leon P. Alford proposed that the ASME review not only its code of ethics, but its entire structure and purposes. The resulting Committee on Aims and Organization became the instrument of revolutionary changes, not only in the ASME, but in the entire profession. The ASCE, acting independently, created its own Committee on Development at the same time, followed by the AIME and the AIEE. These four committees proposed that the Engineering Council be replaced by a more representative body.[70]

Cooke, however, was relegated to the status of a bystander. He was not a member of the Committee on Aims and Organization. And while the committee recommended a new code of ethics and other important reforms of the ASME, these changes were not wholly in accord with Cooke's own views of what ought to be done. The new code of ethics, for example, did make the public interest the engineer's first professional obligation. But it was so vague and general that it was scarcely a code of ethics at all. It was unenforceable, since it forbade nothing. It was not the powerful instrument of professionalism that Cooke had intended to use to drive out the quacks and break the alliance of engineering with the private utilities. [71]

Although Cooke's influence remained significant for several years longer, he was clearly a declining force in ASME affairs. In 1918, the society elected Mortimer E. Cooley as its president. He had been one of the four men Cooke publically attacked as being biased in favor of the utilities. His election suggested that while the membership wanted change, many of them resented Cooke's tactics. Cooke tended to lump together all of those associated with the utilities. This lack of discrimination was probably an error; Cooley was markedly more liberal than men like Jackson and Humphreys. As president, Cooley continued the gradual democratization of the society initiated by his predecessors. Unlike the other engineers attacked by Cooke, Cooley showed no personal vindictiveness and treated Cooke's charges as something of a joke. Cooley's transparent honesty and moderation while president of the ASME probably served further to undermine not only Cooke's influence, but his credibility as well. [72]

Cooke had his revenge, however. During Cooley's presidency, the council gave Cooke another opportunity to attack the alliance between business and engineering. It made the ASME a member of the National Industrial Conference Board, an organization of businessmen. Cooke was by this time a master at mounting a political campaign. He bombarded the secretary of the ASME with protests. He then sent copies of this correspondence to the local sections. The secretary and other officials attempted to placate the sections, but, as Cooke well knew, they were in a mood for rebellion. The council was forced to back down as gracefully as it could. It appointed a committee to investigate. But the result was a foregone conclusion, since the head of the committee was Fred J. Miller, a stalwart reformer and a friend of Cooke's. After due deliberation, the ASME withdrew from the conference board. [73]

Cooke had miscalculated. Although the society continued its course of reform, it turned to more moderate leaders. The members agreed, in large measure, with Cooke's diagnosis of the society's ills, but they were not willing to accept the cure he prescribed. As President Hollis pointed out, an attempt to act in the manner suggested by Cooke would have the effect of dividing the ASME into warring cliques that would mutually destroy each other. [74] Ordinary engineers did not see how public service would end their problems; it seemed to beg the essential problem of bureaucracy. In a sense these criticisms were aside from the point. The foundation of Cooke's program was his belief in the inevitable triumph of scientific management. This, of course, implied a massive restructuring of industrial bureaucracies and a reorganization of the utilities. If these things had come to pass, then Cooke's proposals would have made good sense. But an attempt to apply Cooke's heroic remedy without the concurrent acceptance of scientific management throughout industry would have been plainly disastrous. If Cooke failed, it was because scientific management failed.

Although the mechanical engineers turned to others for leadership, Cooke had had a greater influence on his profession than any other reformer. He demonstrated the potentialities that scientific management had as an ideology for engineers. He showed how, through it, engineering progressivism could be linked with the broader traditions of American reform. Almost singlehandedly, he organized a revolt against the existing leadership of the ASME. With the assistance of colleagues in scientific management, like Miller and Alford, he set the society on a course of democratization. It was Cooke who raised the issue of social responsibility. His Committee on Public Relations was the germ out of which grew the Engineering Council and its more democratic successor, the Federated American Engineering Societies. Despite his waning influence, Cooke was able, along with other members of the scientific-management movement, to contribute in important ways to the formation and the early policies of the FAES. This was appropriate, for Cooke was the leading engineering progressive, and the FAES was the most ambitious attempt by the profession to express a progressive interpretation of the engineer's social role and responsibilities.

NOTES

1. Monte A. Calvert, *The Mechanical Engineer in America, 1830–1910* (Baltimore, 1967), has discussed the transformation of mechanical engineering with emphasis on the split over apprenticeship versus college training.

2. On the importance of New Yorkers and their business connections, see Cooke to Calvin Rice, January 15, 1915, and Rice to Cooke, January 21, 1915, file "ASME—Commercial Control," box 168, Papers of Morris L. Cooke, Franklin D. Roosevelt Library, Hyde Park, New York.

3. Taylor to Cooke, October 6, 1910, file "Cooke, June–December, 1910." See also Taylor to Cooke, December 13, 1909, file "Cooke, 1909," Frederick

William Taylor Collection, Stevens Institute of Technology, Hoboken, New Jersey.

4. "The Boston Meeting of the American Society of Mechanical Engineers," *Engineering News*, XLVII (June 5, 1902), 460–461. There is an extensive correspondence on Taylor's reform of the ASME in the files "ASME Correspondence," for the years 1906 and 1907. See also Taylor to Cooke, October 1, 1906, file "Cooke, 1903–1906," and Cooke to Taylor, July 3, 1910, file "Cooke, 1910," Taylor Papers.

5. Taylor to F. J. Miller, January 7, 1909, file "ASME, 1909–1911," Taylor Papers.

6. Gantt to Taylor, January 16, 1911, and Gantt to Taylor, May 13, 1911, file "Gantt, 1911–1915," Taylor Papers.

7. Taylor to E. D. Meier, April 17, 1911, file "ASME, 1911," Taylor Papers. See also Milton K. Nadworny, "The Society for the Promotion of the Science of Management," *Explorations in Entrepreneurial History*, V (May 15, 1953), 244–247.

8. Kenneth E. Trombley, *The Life and Times of a Happy Liberal, A Biography of Morris Llewellyn Cooke* (New York, 1954), 1–8, *passim*. See also Cooke to Taylor, June 24, 1903, file "Cooke 1903–1909," Taylor Papers.

9. Taylor to Cooke, March 29, 1913, file "Cooke, 1913," Taylor Papers.

10. The adoption of the name has often been ascribed to Louis D. Brandeis, and his use of the term helped fix it upon the popular imagination. But more than six months before Brandeis's action, Cooke persuaded Taylor to use "scientific management" in the title of his *magnum opus*. See Cooke to Taylor, March 27, 1910, file "Cooke, January–May, 1910," Taylor Papers.

11. Cooke to Taylor, October 3, 1910, file "Cooke, January–December, 1910" Taylor Papers.

12. Trombley, *Life and Times*, xiii–xvi, 1–3, 15–19. Taylor's tendency to depreciate Gantt and his hostility toward Gilbreth appear to have owed something to class feeling. Taylor thought Gilbreth "so emphatic in his way that he carries weight with a certain class of men." Cooke commented to Taylor that the title of one of Gantt's books "does not seem to me to be quite dignified" (Taylor to Cooke, September 26, 1910, file "Cooke, June–December, 1910," and Cooke to Taylor, April 30, 1910, file "Cooke, January–May, 1910," Taylor Papers).

13. Samuel Haber, *Efficiency and Uplift, Scientific Management in the Progressive Era, 1890–1920* (Chicago and London, 1964), 99–110, has pointed out the continuity of mugwump ideas and scientific management, emphasizing the themes of nonpartisanship, the strong executive, and the separation of politics from administration. For Cooke's views on the role of "the educated" and

"pseudo-democracy," see M. L. Cooke, *Our Cities Awake* (New York, 1918), 66–67.

14. Cooke to Taylor, February 23, 1909, file "Cooke, 1909," Taylor Papers.

15. Cooke to Taylor, September 25, 1907, and "Memorandum of Some Reasons why Fred J. Miller . . . Should be Re-Elected to Council as Vice-President," file "Cooke, 1907," Taylor Papers.

16. M. L. Cooke, "The Engineer and the People," *Journal ASME*, XXX (October, 1908), 1205–14.

17. Cooke to Taylor, October 24, 1908; October 27, 1908; and November 4, 1908, file "Cooke, 1908," Taylor Papers.

18. Cooke to Taylor, January 12, 1909, file "Cooke, 1909," Taylor Papers; "The Engineer and the People, Discussion," *Journal ASME*, XXXI (March, 1909), 362–380.

19. Cooke to Calvin W. Rice, December 17, 1923, file 144, box 15, Cooke Papers.

20. Cooke to Calvin W. Rice, December 31, 1923, file 144, box 15, Cooke Papers.

21. Cooke to A. G. Christie, June 9, 1921, file 236, box 22, Cooke Papers.

22. "The Smoke Nuisance, A Phase of the Conservation Problem," file "Cooke, 1908," Taylor Papers.

23. Petition to Meetings Committee, April 30, 1909; W. W. Hall to Cooke, May 13, 1909, file "ASME—Commercial Control," box 168, Cooke Papers; and Taylor to Cooke, December 13, 1909, file "Cooke, 1909," Taylor Papers. Cooke revived the issue of smoke abatement when he was on the ASME's Council, but the proposal was vetoed year after year by council members allied with the utilities (personal interview with M. L. Cooke, June 27–28, 1957).

24. Taylor to Cooke, March 3, 1909, file "Cooke, 1909," Taylor Papers.

25. Taylor to Cooke, July 23, 1909, file "Cooke, 1909," Taylor Papers.

26. Cooke to Taylor, July 26, 1909, and Cooke to Taylor, September 22, 1909, file "Cooke, 1909"; Taylor to Gantt, June 14, 1909, and Taylor to Gantt, July 6, 1909, file "Gantt, 1909–1910"; and Gantt to Taylor, June 20, 1909, and Gantt to Taylor, July 19, 1909, file "Gantt, 1909–1910," Taylor Papers.

27. The best source of information on Cooke's administration is his remarkable report, which aroused national attention as much for its style as its contents (Department of Public Works, Philadelphia, *Plain Talk, Report of Director,*

Department of Public Works [Philadelphia, 1914], 1–4, 27, 40, 49, 73–77, 100–101). See also Trombley, *Life and Times*, 14–29.

28. Department of Public Works, Philadelphia, *Annual Report of the Department of Public Works of the City of Philadelphia for the Year Ending December 31, 1915* (Philadelphia, 1916), 5.

29. *Ibid.*, 4.

30. *Ibid.*, 19–21, and Department of Public Works, *Plain Talk*, 7–28, *passim*.

31. Department of Public Works, *Plain Talk*, 1–7, 50, 107–108.

32. *Ibid.*, 62–68, 102–108.

33. William D. Lewis, "The Philadelphia Electric Case," *The Utilities Magazine*, I (May, 1916), 4–12, is a detailed history by Cooke's lawyer. See also Trombley, *Life and Times*, 36–45.

34. *Ibid.*

35. For further comments on Jackson's tactics see Morris L. Cooke, "Opening Remarks," *Utilities Magazine*, I (January, 1916), 4–5.

36. Philadelphia *Public Ledger*, July 18, 1914, clipping in file "Public Utilities Bureau," box NMF 70, Papers of Louis D. Brandeis, School of Law Library, University of Louisville, Louisville, Kentucky.

37. Philadelphia *North American*, July 24, 1914, and November 14, 1914, clippings in file "Public Utilities Bureau," box NMF 70, Brandeis Papers.

38. Cooke to Brandeis, June 10, 1915, and Cooke to Brandeis, July 28, 1915, and "Memorandum as to the Meeting of the Trustees of the Utilities Bureau," December 30, 1914 (typescript), file "Public Utilities Bureau," box NMF 70, Brandeis Papers.

39. "Proceedings of the Conference on Valuation Held in Philadelphia, November 10th to 13th, 1915," *The Utilities Magazine*, I (January, 1916), 1–227.

40. M. L. Cooke, "Some Factors in Municipal Engineering," *Journal ASME*, XXXVII (February, 1915), 82.

41. Frederick W. Ballard, "The Design and Operation of the Cleveland Municipal Electric Light Plant," *ibid.*, 104–108.

42. "Public Service at the Annual Meeting," *Journal ASME*, XXXVI (August, 1914), iii; "Report of the Public Relations Committee," *Journal ASME*, XXXVII (December, 1915), xvi; M. L. Cooke, *How About It?* (Philadelphia, 1917), 12–13;

Frederick R. Hutton, A History of the American Society of Mechanical Engineers, (New York, 1915), 72–73; and Cooke to Taylor, October 21, 1914, and Cooke to Taylor, October 28, 1914, file "Cooke, July–December, 1914," Taylor Papers.

43. Alexander C. Humphreys, "Some Factors in Municipal Engineering, Discussion," Journal ASME, XXXVII (February, 1915), 85.

44. Cooke to Taylor, October 21, 1914 (two letters), Cooke to Taylor, October 28, 1914, Cooke to Taylor, October 30, 1914 (enclosing a copy of the opposition circular), and "To the Members of the American Society of Mechanical Engineers," October 28, 1914, file "Cooke, July–December, 1914"; Cooke to Carl Barth, October 23, 1914, file "Barth, 1913–1915"; and A. C. Humphreys to Taylor and F. J. Miller, March 15, 1915, file "Stevens Institute of Technology," Taylor Papers.

45. Cooke to Carl Barth, October 23, 1914, file "Barth, 1913–1915," Taylor Papers.

46. M. L. Cooke, Snapping Cords (Philadelphia, 1915), 16, 35, 39. Jackson's firm employed some twenty engineers (see "The Presidents of the Four National Engineering Societies," Engineering News, LXV [Feb. 2, 1911], 135–136).

47. See, for example, A. C. Humphreys, "The Present Opportunities and Consequent Responsibilities of the Engineer," Journal ASME, XXXV (January, 1913), 41–71, and Dugald C. Jackson, "Electrical Engineers and the Public," Trans AIEE, XXX (April–June, 1911), 1135–42.

48. Cooke to Calvin W. Rice, January 15, 1915, file "A.S.M.E.—Commercial Control," box 168, Cooke Papers.

49. Cooke to Calvin W. Rice, June 23, 1916, file "A.S.M.E.—Commercial Control," box 168, Cooke Papers; William Kent, "D. S. Jacobus; American Society of Mechanical Engineers," Engineering News, LXXV (February 3, 1916), 219; Cooke, How About It? 13.

50. Cooke, Snapping Cords, 27, and M. L. Cooke, "Suggested By Our Correspondence," The Utilities Magazine, II (March, 1917), 17–18.

51. M. L. Cooke, "The Nationalization of the Private Utility Interests," The Utilities Magazine, I (November, 1916), 12–18.

52. M. L. Cooke, "An Open Letter to the Engineering Profession," The Utilities Magazine, I (July, 1916), 1–3, and M. L. Cooke, "John Eschleman, Railroad Commissioner," The Utilities Magazine, I (May, 1916), 1–3.

53. M. L. Cooke, "Suggested By Our Correspondence," The Utilities Magazine, II (March, 1917), 21.

54. *Ibid.*, 18–19.

55. *Ibid.*, 21.

56. *Ibid.*

57. *Ibid.*, and Cooke, *Snapping Cords*, 22–42.

58. Cooke, *Snapping Cords*, 3, 5–10, 12–17, 21, 23–30, 35, 40.

59. Cooley, Humphreys, Jackson, and Swain to the Council of the ASME, October 6, 1915, printed in Cooke, *How About It?* 29–34. There is a copy of the original letter in box NMF 70 of the Brandeis Papers.

60. M. L. Cooke, "Suggested By Our Correspondence," *The Utilities Magazine*, II (March, 1917), 21.

61. Cooke to Brandeis, October 19, 1915, file "Public Utilities Bureau," box NMF 70, Brandeis Papers.

62. Cooke to Charles Whiting Baker (undated), file "Public Utilities Bureau," box NMF 70, Brandeis Papers; Cooke, *How About It?* 34–43.

63. Cooke, *How About It? passim.*

64. Ira N. Hollis, "Activities of the Society for 1917," *Journal ASME*, XL (January, 1918), 40–43.

65. Mortimer E. Cooley, "President Cooley's Address," *Mechanical Engineering*, XLII (January, 1920), 43–47.

66. Cooke, *How About It?* 14, and Ira N. Hollis, "Engineering and Cooperation," *Journal ASME*, XXXIX (November, 1917), 933–934.

67. M. L. Cooke, "The N.E.L.A. Vernacular," *The Utilities Magazine*, I (September, 1916), 16.

68. Jesse F. Orton to Cooke, August 11, 1917, file "ASME Tenancy in Engineering Societies Building," box 168, Cooke Papers.

69. M. L. Cooke, "The Public Interest as the Bed Rock of Professional Practice," *Journal ASME*, XL (May, 1918), 382–383.

70. L. P. Alford, "The Public Interest as the Bed Rock of Professional Practice, Discussion," *Journal ASME*, XL (September, 1918), 753; L. C. Marburg, "Aims and Organization of the Society," *Mechanical Engineering*, XLI (January, 1919), 12–15.

71. "Report of Special Committee on Code of Ethics," *Mechanical Engineering*, XLII (September, 1920), pt. 2, pp. 123–124.

72. Mortimer E. Cooley, *Scientific Blacksmith* (New York, 1947), 155–160, 241–242, 246–257, suggests that Cooley was a moderate progressive rather than a reactionary.

73. Cooke's strategy may be reconstructed from his correspondence, circulars, and other material in file 232, box 22, of his papers. See also Cooke to Calvin Rice, October 23, 1919, October 29, 1919, and November 29, 1919, file 144, box 15, Cooke Papers. For a contemporary view of the controversy, see "Mechanical Society's Alliance with Employers," *Engineering News-Record*, LXXXIII (October 23, 1919), 750–751.

74. Ira N. Hollis, "Activities of the Society for 1917," *Journal ASME*, XL (January, 1918), 40. Cooke's friend, Fred J. Miller, as president of the ASME in 1920, rejected political activism for the same reasons Hollis had cited (see Fred J. Miller, "The Engineer's Relations to Public Questions," *Mechanical Engineering*, XLII [June, 1920], 336).

8 ■ "THE ENGI- NEERING METHOD PERSONIFIED":

HERBERT HOOVER AND THE FEDERATED AMERICAN ENGINEERING SOCIETIES

In 1919, the engineering profession found an effective leader who could knit together the diverse strands of the engineers' thinking and combine them in a practical program of social action. Herbert Hoover combined technical excellence, professional dedication, and eminent public service in a highly personal blend that appealed to virtually all engineering factions. Conservatives rallied to his praise of free enterprise and traditional individualism. Progressives were delighted by his dedication to reform and social planning. Morris L. Cooke gave Hoover what was, for him, the ultimate accolade; he called him "the engineering method personified."[1]

Hoover's appearance on the engineering scene was well timed. An atmosphere of "great expectations" pervaded the engineering fraternity, a peculiar blend of dread and hope. Radicalism and class conflict aroused fears that were enhanced by the Bolshevik revolution in Russia. Inflationary pressures on engineers themselves produced a militance that made action imperative. At the same time, engineers were optimistic. If social unrest appeared to make inevitable a major reconstruction of society, the example of wartime planning seemed to offer a type of solution that might be highly beneficial for the engineering profession. Engineers had played a leading role in the battle of production; if central planning were to continue and expand after the war, the status and power of engineers might improve dramatically. Hoover gave voice to these aspirations. To many of his colleagues he appeared to be a Moses who could lead both the nation and his profession out of the wilderness.

179

The hopes and fears of engineers enhanced the influence of the progressives. They promised to prepare the profession to exploit its new opportunities and to defend its interests. In all of the founder societies, reform factions had, by 1918, increased power. The ASME, due to Cooke's agitation, was in the lead. In the AIME the progressives appeared victorious after a decade of struggle. The election of Herbert Hoover as president, in 1919, seemed to put the final seal on their hardwon conquest. For a short time, at least, the AIME became one of the most progressive of the founder societies and a staunch supporter of professional unification. The reformers were weakest in the AIEE, but they rallied to the support of the postwar movement. It was, however, the ASCE that underwent the most spectacular change. From 1918 through 1920, a progressive rebellion came close to overthrowing the conservative oligarchy that had long ruled the society.

Two related issues divided the ASCE. One was the control of the society by its New York members. The second was the apparent indifference of the dominant group to the larger social role for the engineer, which seemed to be in the offing in the postwar period. The opposition drew its strength from the local associations of members. These associations had been growing in numbers and vitality for some years. The fact that they had not been authorized by the society's constitution probably worked to their advantage; they were free to act as they chose. Many had allied themselves with local engineering societies that had become centers of political agitation and professional ferment. To check the growing independence of these local groups, the board of directors proposed a series of constitutional amendments that would have increased its powers to deal with dissenters. The associations were to be placed under the control of the board. It would have the power to expel members without a hearing; it would be able to modify the bylaws without previous notice; and the powers of the board would be vested in a small executive committee of six, which could act for the board when it was not in session.[2]

The board's proposals met with immediate opposition from the local sections. The Utah, Colorado, and Northwestern local associations disapproved and made their dissent public. The board countered by passing a resolution censuring these sections for publicizing the society's internal affairs. At the society's annual convention in January, 1918, two leading progressives, Nelson P. Lewis and Gardner S. Williams, denounced the board's amendments. The Cleveland local association went further: its members adopted a resolution calling for the local associations to send delegates to a "Joint Association Constitutional

Conference," which would consider the "desirability and substance of further amendments, particularly with reference to a more democratic management of the Society."[3] The results of this conference would then be reported to the local sections. In effect, the Cleveland members were calling for a revolutionary action by the local associations that would by-pass the official government of the society.

The Clevelanders' proposal aroused widespread support among the members of the ASCE. Resolutions in support of this idea were passed by local associations in Colorado, Duluth, Pittsburgh, Portland, Texas, and elsewhere. The Cleveland group circularized the sections and reported that a majority supported the idea of a constitutional conference. Two of the associations further suggested that a portion of the annual dues be remitted to the local sections.[4] Something of the spirit of the revolt against the ASCE's central government was suggested by F. Herbert Snow, president of the Philadelphia association. He wrote:

> Although hostilities in Europe are over, warfare against waste and inefficiency has only begun. . . . The signs of the times portend an extended exercise of governmental functions along comprehensive, economic lines for the general welfare. . . . Therefore, to not a few observers has come the belief that to the Engineering Profession is about to be accorded its proper place in the economic life of nations, and the Profession is bound to assume a position of the utmost importance in all further activities.[5]

Faced with a revolt of the membership against its authority, the board of directors capitulated. On June 18, 1918, the board resolved to create a Committee on Development. In make-up this committee approximated the constitutional conference proposed by the Clevelanders. Twenty-two of the members were elected by the local associations; only seven were appointed by the ASCE's central government. No restrictions were placed on the committee's activities. It could, and did, range over every aspect of the society's internal organization and external relations. The preamble to the resolution setting up the committee suggested that the directors were not wholly insensible to the postwar opportunities of their profession. It noted that "sociological and economic conditions are in a state of flux," and that in the readjustments that impended the engineer would be in a position to assume a "larger sphere of influence and usefulness." The purpose of the committee was to prepare the ASCE for its enlarged functions.[6]

The formation of the ASCE's Committee on Development aroused intense interest on the part of the entire profession. "Surely this action must mark the end of the ultra-conservative regime," commented *Engineering News*.[7] But the editors thought that the committee's significance greatly transcended the purely domestic affairs of civil engineers. Taken together with the parallel action of the ASME in setting up a Committee on Aims and Organization, the ASCE's decision pointed to a new and important social role for the engineer. The editors were convinced that the existing rulers of the American economy, the bankers, had failed. As a result "the industrial structure is to be recast." The one sure way of solving the nation's problems was by a "scientific study" of all the factors involved; the editors were confident that the new structure would have to be built on engineering foundations. Under these circumstances it appeared likely that the engineer would be one of the principal architects of the new order.[8]

The anticipation of a major reconstruction of society was widely shared within the engineering profession. Clearly, if the profession was to take advantage of its opportunity, it would not be enough for the separate societies simply to modify their own internal organization. Unity by the entire profession was essential. The ASME and ASCE both created their committees in June, 1918. Although the ASME was slightly prior in time, the actions were independent; both represented the culmination of years of progressive agitation. These two committees soon made contact. They approached the AIME and AIEE and persuaded them to set up similar committees. The AIEE, the last to act, created its Committee on Development in January, 1919.[9]

The appearance of the four committees led to two separate but closely related chains of events. Each of the committees considered the changes necessary for its own society. But at the same time, each committee reported to its parent society that the foremost need of the day was increased cooperation among engineers. With the consent of the founder societies, the four committees then met together to form a Joint Conference Committee to consider the means for achieving professional unity. Whether acting individually or collectively, the chief issue for the committees was democracy. There was a general agreement that both the separate societies and the new unity organization should be representative of their members. It was the task of the committees to determine how to achieve this, both for the individual societies and for the organization that was to replace the Engineering Council.[10]

All four committees agreed that new opportunities were opening up for the engineering profession. They assumed that many wartime changes would be permanent and that society would undergo a major postwar reconstruction. They thought that engineers had a duty to assist in the rebuilding. They were not unmindful of the possibility that the profession might better its position in this process. They believed that engineering could and should be used in solving social problems. Something of the hopes of engineers at this time was suggested by the chairman of the ASME's Committee on Aims and Organization. He noted that "these are days of change; indeed of revolution." He predicted that the engineer's "habit of thought" would carry society "far into the promised land of economic efficiency and social justice."[11]

But engineers lacked consensus as to how their profession should take advantage of its opportunity. There was general agreement that the scope of engineering societies should be broadened to include matters of public policy related to engineering and that the Engineering Council should be replaced by a more representative body. But progressives tended to assume that fundamental alterations in the existing pattern of engineering-society government were necessary. They wanted to change the methods of nominating and electing officers. Conservatives tended to hold that the existing constitutions were sufficiently democratic. There was agreement that the Engineering Council's replacement should represent the entire profession and not just the founder societies. But engineers differed on the method of achieving this goal. Should the new organization represent engineers as individuals or should it be composed of societies? If the latter, should the delegates be elected or appointed? On these two questions the division was not just between progressives and conservatives; the reformers were themselves divided.

From 1919 to 1920, the debate on the broad issues of internal democracy and the make-up of the new unity organization centered in the ASME and ASCE. The reformers in the AIME and the AIEE were moderates; they were unwilling to challenge the conservatives. The civil and mechanical engineers provided the militant progressive leadership. It was unfortunate for the progressives that action on these issues required constitutional changes and the cooperation of four separate societies. The reform spirit among engineers was at its height in 1918 and early 1919. But the several committees did not report until 1919, and the Engineering Council's replacement was not organized until the end of 1920. The members of the ASCE did not finally vote on the issues of internal reform and professional unity until the fall of 1920, more than two years after the creation of the Committee on Development. By that time the first rush of enthusiasm was over; had the vote been held earlier, the results might have been more favorable for the progressives.

One of the most important battles between the progressives and conservatives took place in the ASCE. The source of controversy was the report of the Committee on Development. It proved to be more progressive than the insurgents anticipated; it probably was more reformist than the directors had expected when they created the committee. The report suggested a drastic restructuring of the society along the lines long advocated by the progressive opposition. It proposed that all of the society's members be grouped into local sections and that these become the basic governing units of the ASCE. The sections would elect the directors, they would pick the nominating committee, and they would be allotted a regular percentage of the dues to finance their operations. The committee also suggested that the ASCE adopt new membership standards and a new code of ethics. The committee's recommendations, in the form of nine resolutions, were submitted to the ASCE's annual meeting early in 1920. [12]

A few progressives were unhappy with the Committee on Development. Some reformers suspected that the creation of the committee was a delaying tactic on the part of the conservatives. [13] Others were critical of the substance of the report, which they thought did not go far enough. One member was distressed because businessmen and other nonprofessionals would continue to be members of the ASCE. He also urged the society to take stronger steps to protect consulting practice by civil engineers. [14] Another member pointed out that the changed method of selecting the nominating committee left the basic electoral system unchanged. So long as the nominating committee selected only a single candidate for each office and kept its own proceedings shrouded in secrecy, then voting for officers would remain a mere formality. He wanted contested elections to introduce real democracy; he suggested that the nominating committee be abolished altogether. [15]

But most progressives were in favor of the committee's recommendations. Its resolutions promised to break the power of the New York oligarchy and make the society responsive to the membership all across the country. Richard L. Humphrey organized a campaign to secure support for the committee's resolutions. Teams headed by a locally prominent engineer were set up in each region. [16] It was the conservatives, however, who did not accept the recommendations. Early in 1920, a conservative group won control of the New York local section, which became the center of the campaign to defeat the development committee's proposals. [17]

The progressives expected victory in the ASCE; they constituted a clear majority of the membership. But the reform cause went down to defeat. A major reason for this unexpected result was a split among the progressives themselves. Frederick H. Newell and others affiliated with the American Association of Engineers lined up with the conservatives in opposition to the proposal that the ASCE join the new federation of the engineering profession. Newell and his allies thought, quite correctly, that this organization would undermine the AAE. The first round of the battle was won by the proponents of unity. At the society's annual meeting in early 1920, the Committee on Development's recommendations were adopted and submitted to the membership for a letter ballot. In this vote, however, the alliance of the Newell forces with the conservatives was sufficient to defeat the proposal that the ASCE join the new engineering federation.[18] The supporters of federation did not accept this result as final, however, and another ballot was scheduled for the fall of 1920.

The development committee's suggestions for the democratization of the ASCE's government were also defeated. Unlike the issue of joining the new unity organization, the governmental changes necessitated constitutional amendments that required a two-thirds vote for adoption. These amendments were submitted to the membership in a second letter ballot. In this case the Newell supporters did not actively oppose the reforms, but their indifference may have influenced the outcome. All of the amendments got more than a majority of the votes. But the more important ones failed to receive the necessary two-thirds, and they were defeated. The conservatives had won, but only by exploiting internal dissensions among the progressives and by virtue of a constitution that frustrated the will of the majority.[19]

The ASME's Committee on Aims and Organization was far less radical than the ASCE's Committee on Development. The make-up of the two committees was almost identical; both were composed of representatives from each of their local sections, along with a few appointed members. The moderation of the Committee on Aims and Organization was due to the fact that the ASME had already instituted significant measures of decentralization and democratization. Perhaps of even more importance was the fact that Cooke's attacks had led to more conciliatory policies in the administration of the existing constitution. Apart from a new code of ethics, the committee did not recommend that the ASME adopt major internal changes.[20]

Cooke was one of those who disagreed with the committee. Although he was pleased with the great gains that had been made, he realized that the system of government was still basically the same. Cooke thought that two fundamental changes were needed to make the ASME truly democratic. The first was complete publicity and an end to secrecy in engineering-society affairs. The second was universal suffrage. The ASME denied junior members the right to vote or hold office. Cooke pointed out that only the ASME and ASCE disfranchised younger members. But in the ASME the juniors constituted 23 percent of the membership, and in the ASCE they totaled only about 5 percent. Perhaps because it was a less drastic change and one that offered a large immediate impact, Cooke chose to push for universal suffrage first.[21]

Cooke initiated his campaign to give votes to junior members less than two months after the report of the Committee on Aims and Organization. As in the case of his plea for a new code of ethics, Cooke avoided sensationalism. He made a sober, factual survey that showed the ASME's policy was out of line with those of other engineering societies. Cooke's suggestion was approved in 1920, at the society's spring meeting; it then had to be adopted as a constitutional amendment. It took almost two years to get final approval, however. There was little opposition to the principle involved, but Cooke's sponsorship itself was enough to arouse hostility and suspicion.[22] Cooke countered by an extensive letter-writing campaign through which he mustered support for his proposal. Although ultimately successful, Cooke's amendment was not adopted until March, 1922, too late to influence the ASME's policies when they were still flexible. By 1922, the discontent of the younger men was almost over, and a conservative reaction was in full swing.[23]

While his amendment was still bogged down in delays, Cooke attacked the problem of secrecy. In 1921, he gave a paper, "On the Organization of an Engineering Society," which was perhaps his frankest statement of his professional philosophy. Cooke maintained that the engineers were going to be given "supreme authority" by popular demand, since only they could solve the nation's problems. In the new, engineered society of the future, the engineers were not only to emerge as dominant leaders, but their associations were to become important, if informal, branches of government. To fulfill these enlarged functions, engineering societies needed to be reorganized. Executive authority should be expanded, board and committee management restricted. But most of all, greater democracy and devotion to the public interest were needed. The most important impediment to membership control, however, did not lie in the formal rules of voting, Cooke held. Rather it was the secrecy that kept the average member ignorant of the real issues. Cooke advocated a policy of complete publicity. "In the absence of studied and widespread and uncompromising publicity," Cooke warned, "such power as is undoubtedly coming to the technological group may become a menace."[24]

Cooke's influence, however, was by now well past its peak. He failed to generate any enthusiasm. Indeed, his references to such figures as Thorstein Veblen, R. H. Tawney, and Wilhelm Ostwald seem to have offended rather than impressed the audience. The chairman of the ASME's Committee on Aims and Organization, L. C. Marburg, rose to issue a "solemn warning" that engineers would lose all of their influence if they got the reputation of being "a set of highbrows."[25] The society took no action on Cooke's proposal for complete publicity.

The issue of democracy arose not only in the separate societies but also in considering the proposed federation of the entire profession. The most democratic form of government would have been that made possible by an organization based on individuals rather than societies. In this case membership control would be direct. The Joint Conference Committee considered and rejected individual membership. It thought, quite correctly, that such a super organization would be opposed by existing societies as a threat to their identities. The committee also feared that not enough engineers would join. Individual memberships would have put the new organization in direct competition with the AAE; there was no guarantee that in this sort of contest the AAE might not prove to have the greater appeal.[26]

Having opted for a federation of societies, the Joint Conference Committee made no effort to ensure that the new organization would be responsive to the rank-and-file of the profession. It left the election of delegates to the new federation to the governing councils of the member societies. Indeed, the committee appears to have been anxious to insulate the new organization from swings in popular opinion among engineers. Delegates were to serve a four-year term, with one-quarter to be elected annually. Morris L. Cooke compared this proposed government to the ASCE's, if that society's council were elected by the boards of the local engineering societies rather than by the membership. When the Joint Conference Committee's report was submitted to the ASME in December, 1919, the society, apparently at Cooke's instigation, voted to amend it to provide for an annual meeting of delegates elected directly by the membership of the constituent societies. This meeting would then elect officers whose terms would not be longer than three years. However, Cooke's proposal was rejected by the boards of the founder societies and by the organizing conference that drew up the constitution of the new federation. In practice, delegates were appointed, just as had been the case with the Engineering Council. This made possible the same sort of indirect manipulation by powerful interest groups that had been so characteristic of the government of the Engineering Council.[27]

On January 23, 1920, the governing boards of the four founder societies met and approved the report of the Joint Conference Committee in its original form. They authorized the committee to call an organizing conference to bring about professional unity. In the interval before this meeting, the committee drew up a draft constitution to present to the assembled engineers.[28] The committee issued invitations to 110 engineering societies "not organized for commercial purposes" to attend a conference at Washington, D. C., early in June, 1920. Voting was to be proportionate to size, one vote for each thousand members, but each society, however small, had at least one vote. Engineers representing seventy-one societies assembled in Washington, D.C., on June 3, forming a sort of constitutional convention for the profession. In general, those attending the conference were moderate progressives. The ASME's delegation was headed by L. P. Alford; Cooke had not been selected to represent the society. The ASCE's representatives included Richard L. Humphrey and Gardner S. Williams, both anti-Newell progressives. Newell was present as a delegate of the AAE, but as events demonstrated, his influence was small.[29]

The most hotly contested issue was the nature of the proposed federation. Gardner S. Williams introduced a resolution providing for membership by societies and not by individuals. Newell and the AAE delegation opposed, which led to two hours of spirited debate. Newell pointed out that the AAE had been first in the field and that the new organization would duplicate its functions. Newell did not seriously expect the delegates to drop their carefully nurtured plans and join the AAE. Rather, he held that the new organization should be based on individuals rather than societies. Newell probably thought that if individuals were given the choice they would prefer to join the AAE. The spokesmen of the AAE made it apparent why they opposed an organization based on societies. They clearly did not trust the founder societies; they regarded them as autocratic and unrepresentative of the younger men.[30]

Newell's only hope for carrying his idea of a society based on individual memberships was to appeal to the smaller engineering societies. The founder societies were committed to society membership; an alliance with the local societies had been a basic strategy of the AAE since Newell was elected president. But if this was his hope, he was doomed to disappointment. The representatives of the smaller societies opposed the idea of individual memberships. They feared that their societies would lose their separate identity in such an organization. Newell was not completely finished. He proposed that the new federation confine itself to furthering the public welfare, leaving the welfare of engineers to the AAE. This proposal also was defeated.[31]

After deciding the fundamental question of federation, the conference went on to adopt a constitution. The organization as a whole was to be termed the "Federated American Engineering Societies." Its governing body was to be the "American Engineering Council." In general, the constitution followed closely the proposals of the Joint Conference Committee. Cooke's suggestions for democratic control were ignored. But at the instance of the representatives of the technical press, the conference did adopt one important amendment. This committed the federation to the "principle of publicity and open meetings." Another progressive feature was the forthright dedication of the new organization to the public interest.[32]

Although there had been bitter disputes over the basis of membership and other issues, there was none on the selection of a president. Herbert Hoover was the unanimous choice. The source of his popularity lay only in part in his national and international stature. Equally important was that Hoover had achieved a remarkable synthesis of the various strands of engineering thought. Perhaps his greatest appeal was to the reformers. They idolized Hoover because he stood for three things: professionalism, progressivism, and the engineering approach to social problems.

Hoover appeared as the embodiment of the professional spirit in engineering. He was one of those engineers who had been consciously attempting to propagate professional ideals. His translation of Agricola's *De Re Metallica*, a classical book for the history of mining, was specifically intended to "strengthen the traditions of one of the most important and least recognized of the world's professions."[33] His activities in Belgian relief, in Wilson's wartime cabinet, and in postwar European reconstruction exemplified the larger role that engineers had long been predicting for their profession. Hoover was outspoken in extolling the superiority of engineering to other professions. Engineering was creative and constructive; it inculcated honesty and objectivity on the part of its practitioners. He could wax eloquent in praising the unique potentialities of his profession for public service. Engineers, he held, were:

> through the nature of their training, used to precise and efficient thought; through the nature of their calling, standing midway in the conflicts between capital and labor; and above all, being in their collective sense independent of any economic or political interest, they comprise a force in the community

absolutely unique in the solution of many national problems. None of these values can be obtained for the country as a whole unless the engineering profession can speak in an organized and comprehensive manner.[34]

Hoover manifested his progressivism in a series of postwar statements. He warned against both radicalism and reaction. But of the two he thought that reaction was the more dangerous, since it "too often fools the people through subtile channels of obstruction and progressive platitudes."[35] At a time when repression was the order of the day, Hoover insisted on the paramount need for social justice. "We shall never remedy justifiable discontent," he maintained, "until we eradicate the misery which the ruthlessness of individualism has imposed upon a minority."[36] He thought that labor, for example, did have legitimate grievances, and he favored at least some form of collective bargaining.[37] He held that government intervention in the economy was necessary to prevent "economic domination by the few."[38] While opposing government ownership of business, he praised government regulation of monopolies, such as the railroads and public utilities.[39]

Hoover prescribed an engineering remedy for the nation's ills. Perhaps the thing that most distinguished the reforms advocated by engineers from those of other progressives was their emphasis on use rather than ownership. Thus, Hoover called for a great national campaign to promote efficiency and to reduce waste. The prime need, Hoover thought, was a sharp increase in production. This would require "a definite national program" for such problems as rail and water transportation, irrigation and conservation, fuel and electric-power development.[40] It was for engineers to point out how such an increase in efficiency could be produced. It was their duty "to visualize the nation as a single industrial organism and to examine its efficiency towards its only real objective—the maximum production."[41]

The basic thrust of engineering reforms, including Hoover's, was conservative. The shift in emphasis from ownership to use permitted the preservation of basic institutions and values. In seeing reform as a means of preserving private property, individualism, and other traditional fundamentals, Hoover was in the mainstream of his profession's thinking on social issues. Hoover was convinced that

there is somewhere to be found a plan of individualism and associational activities that will preserve the initiative, the inventiveness, the individuality, the character of man and yet will enable us to synchronize socially and economically this gigantic machine that we have built out of applied science.

With others of his profession, Hoover thought "there is no one who could make a better contribution to this than the engineer."[42] Hoover's suggestions for solving the nation's postwar problems were clearly in the tradition of engineering progressivism. His stress on efficient use rather than ownership was one indication. Also typical were his utterances on the need for cooperation rather than competition, the common interests of labor and capital, and the need for large-scale national planning. Cooke's enthusiasm was indicative of Hoover's relation to engineering reform. Cooke thought that Hoover's industrial views provided "the only possible basis" for building an "ordered industrial democracy." Cooke endorsed Hoover for the Republican nomination for President of the United States in 1920. "This opportunity to put a rugged, red-blooded, warm-hearted, common-sense, liberty-loving American in the White House," Cooke wrote, "seems well-nigh providential."[43]

Appearances, however, were deceptive. Hoover differed from progressives like Cooke on several fundamental matters. He did not see engineering or science as the primary sources of human progress. When he sought to trace the contributions of mining engineers to social advancement, his stress was on their role in the rise of liberalism rather than on their technological innovations.[44] In a book analyzing the sources of America's greatness, he scarcely mentioned engineering. Instead, Hoover thought that America's success was due to its social philosophy.[45] What America needed most, therefore, was not science applied to society, but an ideology that would restore and strengthen its basic moral principles, "a definite American substitute" for "these disintegrating theories of Europe."[46] He attempted to elaborate American individualism into such an ideology:"Its inspiration is individual initiative. Its stimulus is competition. Its safeguard is education. Its greatest mentor is free speech and voluntary organization for public good."[47]

Hoover insisted that any attempt at reform must fit within the limits determined by the American social philosophy. This implied that the role of government would be strictly limited; it could guide and encourage positive tendencies, but it could not impose a program upon industry.[48] Yet individualism by itself was insufficient; conscious guidance and coordination were necessary to achieve national efficiency and raise production. Hoover thought he had the solution. American individualism had been traditionally tempered by associational activities. "A profound development of our economic system," Hoover observed, "has been the great growth and consolidation of voluntary local and national associations."[49] Hoover thought that the necessary degree of coordination and planning for the economy could be achieved through the cooperation of these great interest groups. "If we could secure this cooperation," Hoover maintained, "we should have provided a new economic system, based neither on the capitalism of Adam Smith nor upon the socialism of Karl Marx."[50]

Hoover's vision of cooperation between economic groups led him to foresee a social role for the engineer quite different from that advocated by progressives like Cooke. The engineer need not draw up a blueprint for the rebuilding of society, nor would he be its new ruler. His contribution, in Hoover's view, was to bring about cooperation between the great economic groups. The engineer would point out, in general terms, what needed to be done. The actual planning and the control would, of necessity, be in the hands of the large interest groups; that is, with business for the most part. Hoover imagined the engineer functioning as a sort of social catalyst, producing action by others. He did not view the engineer as an independent force in national affairs.[51]

Hoover saw the engineer less as an independent professional than as an ally of business. He thought that the great majority of engineers worked in administrative and executive positions in industry. The profession's increasing interest in public affairs was itself a consequence of the transformation of the engineer from a technical advisor to an administrator of industry. The qualities of engineers that fitted them to assist in solving national problems were, according to Hoover, the ones they shared with business executives: skill in advance planning, coordination, and a sense of organization. Hoover placed little emphasis on the scientific side of engineering, except to praise the virtues of quantitative thinking. He did not suggest that the specific content of the engineer's technical knowledge would be useful in dealing with social problems.[52]

Hoover's view of the engineer was shaped by his experiences as a mining engineer. Mining engineers were more closely involved in management than any others, a fact reflected in the membership standards of the AIME. Of all technical fields, it was the least influenced by science. There was no parallel in mining to the virtual merger with physics that had occurred in electrical engineering. In any case, mining was less influenced by purely technical factors than by business considerations. The dominant fact of mining engineering practice, as Hoover had pointed out earlier, was "the vast preponderance of the commercial over the technical in the daily work of the engineer."[53]

Hoover's position presented the apparently anomalous situation of a progressive program joined to a conservative, business-oriented philosophy. Cooke, once his initial enthusiasm had cooled and disillusionment had set in, regarded Hoover's philosophy as a virtual repudiation of professionalism.[54] But Hoover's synthesis was a remarkable achievement. His version of professionalism was both more practical and more consistent than Cooke's. Hoover avoided the chief source of paradox in the thinking of engineers, the confusion of scientific and moral considerations. His program did not require the creation of a new science of society, nor a drastic reconstruction of government and industry. Hoover thus eliminated the large leaven of fantasy and wish-fulfillment that had influenced the thinking of other engineers. But in so doing he also abandoned the dream of a society led, if not dominated, by engineers. This made it possible for Hoover to gain converts for his philosophy outside the engineering fraternity, particularly within the business community. At the same time, Hoover could argue that his approach was more effective in dealing with the engineer's status. He maintained that

> we are lifting the engineering profession from the back room of blueprints to the first order of national importance and national esteem. We are improving the public estimation of every engineer in the community. We are advancing the interest of the engineer as it has never been advanced before.[55]

Hoover's suggestions aroused great enthusiasm on the part of his fellow engineers. He had a positive, progressive program, but one that conservative engineers could also support. He appeared, at least momentarily, to have united all engineering factions. He proposed immediate action of a sort that would have an important impact on public policy. He seemed to be the ideal person to lead a united profession toward a new day. At the first meeting of the FAES in November, 1920, Hoover was elected president. There were no rival candidates and no contest; Hoover was the choice of all.

Hoover lost no time in implementing his program. He suggested that the FAES undertake a study of waste in industry. Although the FAES itself was not fully organized, the Executive Board approved Hoover's proposal on November 20, giving him the authority to appoint the committee. With this investigation Hoover began a national campaign for efficiency that he was to continue as Secretary of Commerce and that was to be a cardinal element in the Republican domestic program in the 1920s. In this way some of the engineers' reform ideas were actually put into practice.

The investigations of waste represented a curious blending of Hoover's philosophy with scientific management. This was perhaps inevitable, since only the Taylorites had a fully developed technique for dealing with industrial efficiency. Hoover had no precise method of operation in mind, and he was, in any case, absorbed in a variety of affairs. He, therefore, delegated the responsibility for nominating committee members and working out a plan of operations to his associate, Edward Eyre Hunt. Hunt was interested in scientific management and had applied for full membership in the ASME. He turned to the followers of scientific management for talent to staff the committee. Out of the total of seventeen men appointed, eleven were Taylorites, including Cooke, Fred J. Miller, Robert B. Wolf, H. V. R. Scheele, and Leon P. Alford. Wolf and Scheele had been on the executive board of the New Machine; Alford was Gantt's biographer. In addition, L. W. Wallace, the executive secretary of the FAES, served on the committee and provided liaison with the parent society. He, too, was associated with scientific management, and had written a book on Gantt's charts. One of the few committee members who were not in some way associated with scientific management was J. Parke Channing, a progressive mining engineer and a friend of Hoover's. In organization and spirit the committee was as much or more influenced by the ideas of scientific management as by those of Hoover.[56]

Hoover employed scientific management not only in the study of waste but also in the FAES's second major undertaking, an investigation of the twelve-hour day in continuous industries. Cooke had been agitating for an engineering study of the two-shift day ever since the steel strike of 1919. In letters to prominent engineers Cooke argued that the issue was "almost exclusively a responsibility of the engineering profession."[57] Cooke did not wait for the engineers to act, however. He undertook his own investigation with the assistance of Horace B. Drury, an economist interested in scientific management. Cooke sought to arouse Hoover's interest.[58] He may have been successful. As originally proposed by Hoover, the study of waste was to include the twelve-hour day. Limitations of time and money forced the committee to eliminate this subject from its program. But once the study of waste was well under way, Hoover persuaded the FAES to undertake an investigation

of the twelve-hour day as a separate project. The FAES authorized
Hoover to appoint a committee of eight to supervise this new project.
Five of his selections were men allied with scientific management who
were then serving on the waste committee: Cooke, Miller, Wallace,
Alford, and Wolf. The actual field work was done by Horace Drury
along with Bradley Stoughton, the former secretary of the AIME.[59]

The alliance of Hoover with the scientific-management group committed the FAES to a progressive view of the engineer's social responsibilities. But before the study of waste was complete, however, there
were a number of indications of an impending reaction. The enthusiasm
of 1919 and 1920 was starting to decline. Conservative and business-
oriented engineers had never really lost their positions on the boards of
directors and committees of the major engineering societies. By 1921,
the founder societies had rejected proposals for drastic internal democratization that might have limited the power of these individuals. Their
influence was apparent in the selection of representatives to the FAES.
Cooke complained bitterly that the ASME's delegation was dominated by
men allied with the utilities.[60] At the same time the power of the founder
societies in the FAES was enhanced by the failure of most of the smaller
engineering societies to join. Only twenty-nine of the more than one
hundred eligible societies joined the federated. This was less than one-
half of those represented at the organizing conference a few months
earlier.[61]

One of the bitterest blows to the FAES and to the progressive cause
generally was the second refusal of the ASCE to join the new organization. The first vote had been held before the organizing conference, and
the proposal to join had been defeated by an alliance of conservatives
with progressives who were sympathetic with the AAE. No doubt the
supporters of the FAES hoped that the opposition would diminish, once
the federated was a reality. The issue was submitted to the membership
a second time in November, 1920. President Arthur P. Davis made a
rousing speech for professional unity. He denounced the conservatives
who, "due to the peculiarities of an antiquated constitution, and to a
particularly effective organization," had been able to dominate the
ASCE in the past. They were, he charged, the same group that had
opposed the ASCE's joining the common headquarters building and
every subsequent step toward cooperation among engineers.[62] The
proposal to join the FAES was, however, defeated by the membership.[63]
The tide of progressive reform in the engineering profession had already
begun to ebb.

NOTES

1. M. L. Cooke, "Public Engineering and Human Progress," *The Journal of the Cleveland Engineering Society,* IX (January, 1917), 252.

2. "Sixty-Fifth Annual Meeting," *Proc ASCE,* XLIV (February, 1918), 105–106, 123–126.

3. "Cleveland Association," *Proc ASCE,* XLIV (April, 1918), 356–357.

4. "Cleveland Association," *Proc ASCE,* XLIV (August, 1918), 624–625.

5. F. Herbert Snow, "The Engineer and the State," *Proc ASCE,* XLV (January, 1919), 11–12.

6. "Society Items of Interest," *Proc ASCE,* XLIV (June, 1918), 604–605.

7. "Committee on Development of the American Society of Civil Engineers," *Engineering News-Record,* LXXX (June 27, 1918), 1209.

8. "Preparing the Profession for a Great Opportunity," *Engineering News-Record,* LXXXI (October 10, 1918), 651.

9. "How the World's Largest Engineering Society Came Into Existence," *Mining and Metallurgy,* I (July, 1920), 5.

10. *Ibid.*

11. L. C. Marburg, "Aims and Organization of the Society," *Mechanical Engineering,* XLI (January, 1919), 12–13. Marburg added the qualification, "not to the millenium," which might be taken as a slap at the more extreme proponents of scientific management, like Cooke, who did anticipate something like a millenium.

12. "Report of the Committee on Development," *Proc ASCE,* XLV (October–December, 1919), 889–925; "Report in Full of the Sixty-Seventh Annual Meeting," *Proc ASCE,* XLVI (February, 1920), 200.

13. "Southern California Association," *Proc ASCE,* XLV (March, 1919), 325–326.

14. "Discussion of the Report of the Committee on Development," *Proc ASCE,* XLV (January, 1920), 11–21.

15. "Letters Relating to the Report of the Committee on Development," *Proc ASCE,* XLV (October–December, 1919), 920–921.

16. "Campaign Begun to Support Development Report," *Engineering News-Record,* LXXXIII (December 25, 1919), 1077–78.

17. "Gossip Regarding New York Section, Am. Soc. C. E.," *Engineering News-Record*, LXXXIV (March 4, 1920), 490, and "Oppose Comprehensive Organization of Engineers," *Engineering News-Record*, LXXXIV (February 26, 1920), 422.

18. F. H. Newell, "Shall the Am. Soc. C. E. Enter the Field of Civic and Economic Affairs," *Engineering News-Record*, LXXXIV (March 11, 1920), 536, and Isham Randolph, "Glad Question Three Was Defeated" (letter to the editors), *Engineering News-Record*, LXXXIV (May 13, 1920), 978.

19. "Am. Soc. C. E. Vote on Amendments Canvassed," *Engineering News-Record*, LXXXV (October 7, 1920), 719, and "Am. Society C. E. Amendments to Constitution Defeated," *Engineering News-Record*, LXXXV (October 14, 1920), 767.

20. Charles T. Main, "Broader Opportunities for the Engineer," *Trans ASME*, XL (1918), 478–479, 483; "Recommendations on the Policy of the A.S.M.E.," *Mechanical Engineering*, XLII (January, 1920), pt. 2, pp. 7–9.

21. "Should A.S.M.E. Juniors Have the Privileges of Members," *Mechanical Engineering*, XLII (March, 1920), pt. 2, p. 45; "Spring Meeting of the American Society of Mechanical Engineers," *Mechanical Engineering*, XLII (July, 1920), 388.

22. F. J. Miller to Cooke, June 14, 1920, file "ASME—Amendment re Juniors," box 169, Papers of Morris L. Cooke, Franklin D. Roosevelt Library, Hyde Park, New York.

23. Margaret R. McKim to F. B. Gilbreth, March 10, 1922, file 0816–70, Papers of Frank B. Gilbreth, Industrial Relations Library, Purdue University, Lafayette, Indiana. There are two extensive folders of Cooke's correspondence on this amendment in box 169 of the Cooke Papers.

24. M. L. Cooke, "On the Organization of an Engineering Society," *Mechanical Engineering* XLIII (May, 1921), 324.

25. "Organization of an Engineering Society Discussed at A.S.M.E. Spring Meeting," *Mechanical Engineering*, XLIII (August, 1921), 539.

26. Calvert Townley, "Proposed Federation of Engineering Societies," *Journal AIEE*, XXXIX (March, 1920), 281

27. M. L. Cooke, "Democratic Representation in Proposed Society Federation" (letter to the editors), *Engineering News-Record*, LXXXIV (February 26, 1920), 439; "Report of the Joint Conference Committee," *Bulletin AIME* (December, 1919), xiv–xix; "Constitution and By-laws of the Federated American Engineering Societies," *Mining and Metallurgy*, I (July, 1920), 10–13.

28. Allen H. Rogers, "National Organization of Engineering Societies," *Mining and Metallurgy*, I (May, 1920), 3.

29. "How the World's Largest Engineering Society Came into Existence," *Mining and Metallurgy,* I (July, 1920), 5–6, and "Organizing Conference Plans Federation of Engineering Societies," *Mechanical Engineering,* XLII (July, 1920), 422–423. The ASCE was represented, despite the adverse vote of the membership, because a second vote on the issue was scheduled.

30. Proceedings, Organizing Conference F.A.E.S., Washington, June 3–4, 1920 (stenographic transcript), 30–40, 53–54, American Engineering Society Papers, Engineering Societies Library, New York, New York.

31. *Ibid.,* 45–49, 90.

32. "Account of the Organization of the Federated American Engineering Societies," *Mining and Metallurgy,* I (July, 1920), 7–10, and "Constitution and By-laws of the Federated American Engineering Societies," *ibid.,* 10–13.

33. Herbert Hoover and Lou H. Hoover, translators' introduction, in Georgius Agricola, *De Re Metallica* (New York, 1950), iii.

34. Hoover to Richard L. Humphrey, February 1, 1923, file "Federated American Engineering Societies, 1922–1924," box 1-I/128, Herbert Hoover Papers, Herbert Hoover Presidential Library, West Branch, Iowa.

35. Herbert Hoover, "The Only Way Out," *Mining and Metallurgy,* I (March, 1920), 7. This and other addresses by Hoover were frequently reprinted, sometimes under different titles, and occasionally in abstract. In the Hoover Papers there is a complete set of his publications in bound volumes, under the title *Addresses, Letters, Magazine Articles, Press Statements, etc.* (hereafter cited as *Addresses*). Listed are the various titles of Hoover's statements and places in which they appeared. This speech is there titled "Inauguration Address—A.I.M.E., Sept. 16, 1919" (*Addresses,* II).

36. Herbert Hoover, "Impressions of Socialism in Europe," *Bulletin AIME* (October, 1919), xvii. This appears as "Address at Dinner Given in his Honor by American Institute of Mining and Metallurgical Engineers, Sept. 16, 1919," in *Addresses,* I, Hoover Papers.

37. Hoover favored company unions, although he did not like that term. See Herbert Hoover, *The Memoirs of Herbert Hoover,* 2 vols. (New York, 1952), II, 30–31.

38. Herbert Hoover, "Impressions of Socialism in Europe," *Bulletin AIME* (October, 1919), xviii.

39. Herbert Hoover, "National Program for Great Engineering Problems," *Mining and Metallurgy,* I (October, 1920), 3 (titled "A.I.M.E. Banquet, Minneapolis,—Address, Aug. 26, 1920," in *Addresses,* IV, Hoover Papers).

40. *Ibid.*, and Samuel P. Hays, *Conservation and the Gospel of Efficiency, the Progressive Conservation Movement, 1890–1920* (Cambridge, 1959), 262–266. Hays was the first modern scholar to stress the importance of the distinction between "use" and "ownership."

41. Herbert Hoover, "Extracts from Address of Herbert Hoover, President American Engineering Council," *Journal AIEE*, XL (March, 1921), 253 (titled "Address Before American Engineering Council's Executive Board and Convention of Engineers at Syracuse, New York, Feb. 14, 1921," in *Addresses*, VI, Hoover Papers).

42. "American Engineering Council, Annual Meeting, Washington, January 10–11, 1924," *Journal AIEE*, XLIII (February, 1924), 165.

43. Cooke to the editor, Philadelphia *Public Ledger*, March 23, 1920, file 253, box 23, Cooke Papers.

44. "The Medal Presentation," *Proceedings of the Mining and Metallurgical Society of America*, VII (June 30, 1914), 94–104.

45. Herbert Hoover, *American Individualism* (New York, 1929), 63–71.

46. Herbert Hoover, "Impressions of Socialism in Europe," *Bulletin AIME* (October, 1919), xvii.

47. Herbert Hoover, "The Only Way Out," *Mining and Metallurgy*, I (March, 1920), 7.

48. *Ibid.*, 7–8.

49. Herbert Hoover, "Great Area of Common Concern Between Engineers, Employers and Employees," *Mining and Metallurgy*, I (December, 1920), 4 (titled "Address of Herbert Hoover Before the Federated American Engineering Societies at Washington, D. C., November 19, 1920," in *Addresses*, V, Hoover Papers).

50. *Mining and Metallurgy*, I (December, 1920), 7.

51. *Ibid.*, 4.

52. Herbert Hoover, "The Only Way Out," *Mining and Metallurgy*, I (March, 1920), 3, 6, 7. See also Herbert Hoover, "The Engineer's Contribution to Modern Life," *Mining and Metallurgy*, IX (March, 1928), 104–105, for a later, but more complete, statement of his view of the engineer's social role.

53. Herbert Hoover, *Principles of Mining* (New York, 1909), 185.

54. Cooke approved of Hoover's early policies as president of the FAES, particularly his study of waste in industry. However, he was disillusioned by Hoover's sympathy for the utilities (interview with Morris L. Cooke, June 27 and 28, 1957).

55. Herbert Hoover, "Engineering Lifted from Back Room of Blueprints to First Order of National Importance," Mining and Metallurgy, II (March, 1921), 18–19 (titled "A.I.M.E. Address, Waldorf Astoria, Feb. 16, 1921," in Addresses, VI, Hoover Papers).

56. There is a wealth of information on the organization, personnel, and policies of the waste committee in boxes 1-I/582, 583, 584, 585, and 586 of the Hoover Papers. See also Edward E. Hunt, "Notes on Economic and Social Surveys," in file "Hunt, E. E., 1923–1928," box 1-I/167, Hoover Papers.

57. Cooke to F. B. Gilbreth, August 28, 1920, file 0816–70, Gilbreth Papers.

58. There is an extensive correspondence on Cooke's and Drury's early work on the twelve-hour day in boxes 179 and 180 of the Cooke Papers. On Cooke's efforts to influence Hoover, see Cooke to Edward E. Hunt, January 11, 1921, and Hunt to Cooke, January 11, 1921, file "Cooke, Morris L.," box 1-I/583, Hoover Papers.

59. Committee on Work Periods in Continuous Industry, The Twelve-Hour Shift in Industry (New York, 1922), 3–4.

60. Cooke to Calvin W. Rice, September 7, 1921, file 144, box 15, Cooke Papers.

61. "History and Review of Work of Federated American Engineering Societies," Bulletin of the Federated American Engineering Societies, II (February–March, 1923), 3.

62. Arthur P. Davis, "Federated Engineering Societies of America," Trans ASCE, LXXXIV (1921), 137. Davis apparently had forgotten that he had been among those who opposed the common headquarters building (see "Opinions of Washington Members of the American Society of Civil Engineers Respecting the Union Engineering Building," Engineering News, LI [Feb. 11, 1904], 131).

63. "Federation Decisively Defeated," Engineering News-Record, LXXXV (November 11, 1920), 917.

9 ▪ THE RETURN
TO NORMALCY,
1921-1929

The formation of the Federated American Engineering Societies marked the high tide of engineering progressivism. Hoover and his supporters in the FAES attempted to put their reform ideas into practice by studies of waste and of the twelve-hour day in industry. These investigations produced a conservative reaction that led to the defeat of progressive leaders throughout the profession. This change paralleled the nation's return to "normalcy" and the election of Warren G. Harding. As America lost interest in reform, engineers either rejected the idea of social responsibility or gave it a conservative interpretation. The close alliance between engineering and business that developed in the 1920s brought many material benefits. But the profession lost much of its precious independence. The studies of waste and the twelve-hour day demonstrated the sort of contributions an autonomous engineering profession might have made to national life.

The Committee on the Elimination of Waste began its operations early in 1921. The plan agreed upon was to make a rapid survey in order to stimulate immediate action. The committee itself was too small to do all the work; therefore, it retained established engineering firms, most of which supplied their services at cost or less. Some fifty engineers each spent approximately two months gathering information on 228 plants in six industries: the metal trades, textile manufacturing, printing, boot and shoe manufacturing, the building industry, and the men's clothing industry. The resulting reports were submitted to other experts

201

for criticism. In all, about eighty engineers had a hand in the investigation.[1]

The study of waste was an attempt to produce major results quickly. It suffered from the pressure of time and shortages of funds. The original plan to survey ten industries was cut to six. In addition, the committee gathered historical and other data from the available literature and their own experience. The appointments to the committee were made in mid-January; a summary of the final report was presented to the American Engineering Council on June 3, 1921, only five months later. A summary of the report was given to the press in June, although the final report was not scheduled for distribution until October.[2]

A major difficulty faced the committee at the very outset: to define and measure waste. The committee used a common-sense approach to waste, defining it as the difference between the average efficiency of the enterprises sampled in each industry and the most efficient plant in the same class. But precise measurement was another matter. The committee worked out no clear guide lines, leaving the actual evaluations to the investigators. The committee adopted a standard questionnaire and evaluation sheet, but this failed to eliminate the difficulty.[3] Mr. Charles E. Knoepple, who furnished much of the material on which the questionnaire was based, refused to use it at all. No two investigators used it the same way. One committee member threw out the data sheets of one of his investigators because they did not fit his preconceived opinions. Another, Harrington Emerson, treated the evaluations as a joke: he made out one rating sheet for all the railroads in the United States, a performance Edward Eyre Hunt labeled as "farcical."[4] The lack of any common guide lines was shown by the fact that different investigators often differed widely in their evaluations of the same plant. The committee's experience suggested that scientific management was something less than an exact science.

The waste study involved gathering and reporting on a vast amount of material. Much of the report consisted of sober analyses of the structure and functioning of the six industries selected, and these separate reports contained a wealth of material on such subjects as intermittent employment, industrial accidents, labor relations, and the like. Four chief causes of waste were delineated: 1) low production due to faulty management of plant, equipment, materials, or men; 2) interrupted production caused by idle men or machines; 3) restricted production intentionally caused by owners, management, or labor; and 4) low production caused by ill health, physical defects, and industrial accidents.[5]

The most controversial conclusions of the committee were those that assessed blame for inefficiency. The committee thought that management was responsible for over half of all waste, but labor was blamed for less than one-quarter. Actually, half was the lowest figure for management, that for the textile industry. Management's share was 75 percent in the men's clothing industry and 81 percent in the metal trades.[6] It was not clear what these figures meant in real terms because of the subjectivity inherent in the committee's methods of evaluation. Two factors probably account for the heavy weight assigned management. One was the general tendency of those in the scientific-management movement to assume that virtually all waste was potentially removable through good management. The other was semantic. Responsibility was defined in terms of opportunity for removing waste, rather than of moral responsibility for causing it.[7] By this token a socialist committee would have assigned the blame for most waste to government, since by this frame of reference, government would be considered the agency best suited to remedy the situation.

The preliminary report of the waste committee aroused a furor when it was presented to the executive board of the Federated American Engineering Societies in June, 1921. Contrary to the FAES's constitution, the debate was conducted in a secret executive session. A substantial number of the members of the executive board were angered by the committee's aspersions against American business management. Others argued that the report was not an engineering document at all. Hoover, though not entirely happy with the result, defended it; and he was apparently instrumental in working out a compromise whereby the report was published simply as the report of the committee and not under the name or with the official endorsement of the FAES.[8] Hoover, by this time in Harding's cabinet, wrote an introduction to the report in which he carefully pointed out that no moral onus was implied by the committee's use of the term "waste." He proceeded to set up a group within the Department of Commerce to carry out some of the recommendations contained in the report.[9]

According to Morris L. Cooke, the waste study was held to be "lèse-majesté" in employer circles, since it blamed them for most of the waste.[10] One influential business organization, the National Industrial Conference Board, organized an extensive campaign to discredit the waste report.[11] Industrial pressure no doubt played a significant role in the reaction the waste report helped trigger within the engineering profession.

The waste committee's report touched an exceedingly sensitive nerve within the membership of the founder societies and set off a reaction against the more progressive idea of social responsibility and against the idea that engineers ought to attempt to influence public policy at all. Only the ASME received the report with official approval. The leadership of the AIME was not prepared for the kind of study presented. In March, 1921, just before the preliminary report, President Edwin Ludlow expressed the wish that the council should cause the engineer to become "a power for maintaining American ideals and standards throughout the land." He thought that "there has never been a time in all our history when sane, conservative influences were more needed."[12] Since the study was sponsored by Herbert Hoover, the AIME made no public criticism. But privately, the board protested the tone of the report as revealed by the press releases issued in June.[13] Although no official denunciation was issued by the institute, W. R. Ingalls wrote a series of editorials highly critical of the waste report.[14] T. A. Rickard and other members objected to the inflammatory, reactionary tone of Ingalls's editorials, but they continued to appear in *Mining and Metallurgy;* there can be little doubt that they were approved by a majority of the institute's board of directors.[15]

The most violent reaction of all appeared in the AIEE. When Arthur W. Berresford, a utilities executive, had been elected president of the institute early in 1921, he had warmly endorsed the idea of civic action by engineers. The engineer, he maintained, was especially fit to serve society because of his altruism, his creativity, and his professional status. He thought that the AIEE's expansion to new fields was both inevitable and desirable.[16] However, shortly after the release of the preliminary report of the waste committee, he took a different view of the engineer's relation to social questions. He commented:

> I think I perceive a growing resentment toward this conception of the engineer's function. I am not at all certain that the engineer should presume to advise in the relation between employer and employee, nor that he should consider himself competent to adjust the many problems which arise in commerce and industry.[17]

Berresford thought that "the major problem of the present day is the education of the mass of the people to the perception of true values." He did not consider the engineer especially qualified for this task; rather he thought that the enunciation of principles should be by the "best minds" of industry who, he felt, were to be found "in executive personnel of our great corporations." He concluded, "I believe that these same leaders of the industry are the most competent to determine the fundamentals and the policies by which we should be guided."[18]

The reaction begun by the waste study was consummated when the FAES completed its next major investigation, a study of the twelve-hour day in the steel industry. This study was technically superior to the earlier investigation of waste. The field work was entrusted to two very able men: Horace Drury, who had had a year's experience in working on the problem, and Bradley Stoughton, a metallurgist and former secretary of the AIME. The atmosphere of feverish haste, so characteristic of the waste study, was absent. The investigation was thorough and painstaking. Methodological pitfalls were avoided. The final report was couched in careful language, and no attempt was made to assess blame against any group.[19] The report, which read like the technical treatise it was, was submitted to the FAES on September 8, 1922. The investigators, in enunciating their guiding principles, hopefully commented that "this declaration the committee believes is not controversial."[20] The report's tone of respectability was enhanced when Hoover drafted an introduction, which he persuaded President Harding to sign.[21]

The committee's hope that it had avoided controversy proved ill-founded. The report clearly indicated that the change from the twelve-hour to the eight-hour day in continuous industries was not only practical but advantageous. In those plants that had made the change, economic loss had been slight or nil, and great improvements had been noted in the workers' morale and the quality of their work. Very few of the plants adopting the three-shift day had subsequently reverted to the two-shift system.[22] However soberly phrased, these conclusions were highly controversial. The steel industry had been fighting bitterly against a shortened day in the face of strong pressure from the White House and public opinion. The steel managers, and their many allies in the engineering profession, could only regard the federated's study as a stab in the back.[23]

The report on the twelve-hour day, perhaps understandably, aroused a greater reaction than the waste report and led to the virtual crippling of the FAES. As with the waste study, the executive board refused to endorse the report, and it was printed simply as a report of the committee.[24] Of the founder societies, only the ASME was favorable. The AIEE ignored the report almost entirely. In the ASCE, the report may have weakened support for the federated. Although initially defeated, the supporters of engineering unity were still strong in the ASCE; but when a new vote was taken after the waste and twelve-hour day studies, the proposal to join the federated was defeated by a larger majority than before. This was widely interpreted as a vote of no confidence in the FAES.[25] Within the AIME, where the steel industry was very well represented, there occurred a more violent reaction. *Mining and Metallurgy* carried no summary of the American Engineering Council's report, but pointedly published the text of a defense of the twelve-hour day prepared by the American Iron and Steel Institute. Editorially, W. R. Ingalls severely attacked the federated's study.[26] Most important of all, the leaders of the AIME decided to withdraw from the FAES.

Although a majority of the AIME's directors were in favor of withdrawal, they had to cope with a sizeable opposition from among the membership; above all, they had to contend with the immense prestige of Herbert Hoover. Therefore, the leaders of the secessionist movement employed a roundabout strategy: they argued that the institute lacked sufficient funds to continue its membership in the federated. A move to withdraw the AIME was not wholly unexpected. A year earlier, in November, 1921, the AIME Board of Directors had threatened to withdraw, apparently in response not only to the waste report but also to the federated's refusal to take a stand against all licensing laws for engineers, a subject on which many mining engineers felt very strongly. This move, however, had been squelched by Hoover and his ally J. Parke Channing.[27] On October 20, 1922, six weeks after the presentation of the twelve-hour day report, the Board of Directors of the AIME concluded that the ordinary funds of the institute could not be used to support the FAES. The board sent a letter to the members of the institute inviting them to contribute to a special fund for this purpose, "in the event of their approval of the institute remaining a member of the federated."[28] This rather qualified "appeal" met with a fairly good response; about 650 members, 10 percent of the total, contributed something over $4,000, enough for the AIME to remain in the federated for six months. Some of the AIME leaders held that this amounted to a referendum and that it indicated only a minority of the membership was interested in the work of the federated. Channing attacked this contention, and he was able to carry the day in an address to the board in June, 1923. A resolution was passed allowing the AIME to remain in the FAES through the end of 1923.[29]

Although Hoover and Channing had been able to delay the AIME's withdrawal, they were unable to prevent it. The board of the AIME refused to approve funds for 1924, and the institute formally withdrew in January, 1924. Channing wrote a bitter letter to the AIME directors in which he made a detailed critique of their claim that they had been motivated solely by financial considerations.[30] To Hoover he lamented that "it is a sad commentary on the engineers of the United States, after having preached cooperation and efficiency, to have them show their utter incapacity for those things themselves."[31] Hoover was no more convinced by the board than Channing. "That the mining engineers of the United States cannot stand a tax of $1.00 a year for purposes of clean cut public service seems to me preposterous," he wrote a friend in a letter later published in *Mining and Metallurgy*.[32] Four of the institute's local sections requested the board to reconsider its position and rejoin, but without success.[33]

The withdrawal of the AIME and the second refusal of the ASCE to join were great blows to the FAES in terms of both prestige and financial resources. Dean Mortimer E. Cooley, Hoover's successor as president, announced his resignation in October, 1923, probably in response to these actions. Cooley was pessimistic about the federated's future, at least in private correspondence. He thought that to succeed the FAES needed the sort of favorable publicity provided by the waste and twelve-hour day reports; but he concluded that further studies of this character were now out of the question.[34] The reason for this, apart from the changing climate of opinion within the profession, was the adoption of a series of constitutional changes that were to cripple the federated for the remainder of its existence. At the same time, the open meetings provided for by the constitution were unofficially abandoned.[35]

The amendments to the constitution gave vetoes to almost every vested interest represented in the FAES. No policy decision could be taken without the assent of two-thirds of each delegation from a national society having six or more representatives. Thus, three or four delegates could block any action. The other societies represented were also given vetoes; all policy decisions had to have the assent of two-thirds of the aggregate of the delegates of the smaller national societies and also two-thirds of all the representatives of the local and regional associations. The vice-president commented that "this provision prevents any hasty or ill-considered action being taken over the opposition of even a comparatively small minority."[36]

It was appropriate that the older name, "Federated American Engineering Societies," was dropped at this time and replaced by "American Engineering Council." For the significance of the reorganization was a reversion to the ideas and policies of the Engineering Council. Like the Engineering Council, the AEC was dominated by business interests who exercised an effective veto over its affairs. This followed not only from the new constitution of the AEC, which made veto power easy to obtain, but also from the character of the delegations selected by the member societies. Although the ASME publically approved of the waste and twelve-hour-day studies, it is significant that six out of nine delegates elected in the latter part of 1921 were affiliated with electrical utilities.[37] Thus, the ASME's representatives alone gave the utilities a veto over any action by the AEC. The new name was appropriate also in that the constitutional changes restored the dominant position of the founder societies, even with the absence of the AIME and ASCE.

The most important way in which the AEC resembled the older Engineering Council was in expressing a conservative, business-oriented view of the engineers' social role and responsibilities. Progressives had emphasized the need for genuine reforms in the business system. The studies of waste and the twelve-hour day suggested how engineers might contribute to social reconstruction. The AEC, however, became a defender of the business system and of *laissez faire*. It stressed the engineer's responsibility to defend correct moral principles. This change in orientation was foreshadowed in September, 1922, when one of the federated's directors, J. C. Ralston, made a plea for the engineer to defend American industrial ideals. He feared that the forces of bolshevism, class-consciousness, and disloyalty, if not arrested, would destroy the nation's social and political ideals. The American industrial ideal, he felt, was "republican" not "democratic," and he urged the FAES to "bring forcefully before all engineers and engineering colleges of the country a clear and sound understanding of what our Industrial Ideal should be."[38] A committee was set up to formulate a plan for presenting this subject to engineering colleges, but nothing was done other than the writing of some innocuous letters.[39]

Although a few progressives were elected to the council, the only policies that could be adopted were those favored by the more conservative wing of the business community. One progressive victory came in 1926 when the AEC approved, at least in principle, a large five-year program of investigations. Included were proposed studies of agricultural waste, waste in specific industries, and an engineering study of the labor problem. The estimated cost was $335,000.[40] But this proved to be a victory on paper only; the AEC took no action on these or any other suggestions that might have antagonized important business interests.

Instead of bold innovation, the AEC came to represent a position of doctrinaire *laissez faire*. The council adopted a resolution in December, 1925, which stated that the national government should not "trespass" in the field of business.[41] That this was no empty gesture was shown by the council's opposition to a favorite plan of Herbert Hoover's, the

Boulder Dam project. In 1928, the council passed resolutions against the expenditure of public funds for the proposed dam at Boulder Canyon on the Colorado "on the ground that it involves Federal ownership and the sale of power."[42] The council also indicated its hostility to federal aid to civil aviation or to agriculture. It objected to attempts to set up a Muscle Shoals Commission.[43]

The negativism of the AEC was all the more unfortunate in that Hoover's tenure as Secretary of Commerce provided a golden opportunity for engineers to influence public policy. Hoover did what progressive engineers had been advocating for two decades; he organized a great campaign for national efficiency. Hoover reorganized the Department of Commerce, giving priority to questions of waste. He was successful in gaining the cooperation of the great business interests of the country. His technique was

> to take up some area where progress was manifestly possible, thoroughly to investigate its technology, and then to convene a preliminary meeting of representatives of that particular segment of industry, business and labor. If the preliminary meeting developed a program, a committee was appointed to cooperate . . . in its advancement.[44]

In all, more than 3,000 such conferences were held during Hoover's tenure as Secretary of Commerce.

The AEC cooperated with the Department of Commerce in a number of enterprises, but the results were, at best, minor. The council was represented on a number of the committees that Hoover organized, including those on business cycles, recent economic changes, and recent social trends. It undertook some new investigations in cooperation with the department in such matters as airport drainage, traffic signs, and coal storage. But in its determination to avoid controversy, the council doomed itself to triviality. The AEC refused to take part in one of the Department of Commerce's major efforts, that of standardization. While supporting the broad movement, the council refused to be involved in the determination of specific standards.[45] Engineers participated in Hoover's campaign for national efficiency, but as auxiliaries, not policymakers.

The AEC's alliance with industrial interests probably accounts for some of the support it continued to receive. The AIEE's loyalty may have been due to the fact that the electrical utilities derived considerable benefit from the council. Two of the AEC's later presidents were utility executives, Arthur Berresford and William S. Lee. In addition to frequent condemnations of Muscle Shoals bills, the AEC was also against measures to encourage rural electrification.[46] The council dropped a proposal to investigate public-utility regulation when certain members expressed their opposition.[47] Possibly one reason that the ASCE finally changed its mind and joined the AEC in 1929 was the support which the council offered the railroads. The council objected to bills designed to outlaw wooden cars on railroads, although it recognized that the use of wooden cars in conjunction with steel ones created a grave danger to passengers. Its actions in this matter were based on the claim that the railroads were eliminating the wooden cars as fast as finances permitted, but no evidence was presented to the council to support this contention.[48]

The AEC's actions on public-policy questions were notable for their lack of factual data and objective reasoning—the very things that engineers thought of as their special contribution to national life. The usual procedure was to refer to dogmatic principles, such as *laissez faire* or states rights, rather than to indulge in fact gathering. In 1931 and 1932, the council opposed bills to limit stream pollution. The council's reasoning was twofold: that no problem existed and federal action was undesirable. The council argued, without presenting any detailed facts, that the federal government was the worst offender, that oil pollution was on the decline, and that the states had the problem under control. It also objected to interference by the federal government in an area of state jurisdiction.[49] On many issues the council took the abstract position that any governmental competition with private enterprise was evil.[50] Lacking imagination, independence, or unity of purpose, the AEC gradually lapsed into a well-merited obscurity.

The AEC's rejection of the progressive concept of the engineer's role and responsibilities paralleled the nation's return to "normalcy." The major engineering societies underwent a similar reaction. Progressive leaders like Cooke and Newell were repudiated by their societies. There was a resurgence in the power of the boards of directors and a decline of democracy in engineering societies. Perhaps the most striking change was ideological. Engineering progressivism underwent an internal collapse; scientific management and conservation, for example, ceased to be forces for engineering reform. Where an active social and political role for engineers was not renounced altogether, it was given a conservative interpretation that stressed adherence to correct moral principles and the alliance or identity of the engineer and the businessman.

Perhaps nothing indicated the decline in the spirit of reform more clearly than the repudiation of Newell and Drayer by the American

Association of Engineers. Founded as a protest organization by younger men, the AAE's position became anomalous when the economic position of younger engineers underwent a marked improvement. After reaching a peak membership of more than twenty thousand in 1921, the association began to decline. By the middle of 1922, more than one-quarter of its members were more than three months in arrears of dues. At the 1921 annual convention, an insurgent faction seized control and abolished Newell's position, that of director of field forces. Only a last-minute rally of Newell's supporters prevented the supercession of Drayer as secretary also. Drayer managed to keep a precarious tenure until 1925, when he too was dismissed.[51] The association continued to decline in membership for some years, finally stabilizing at about five thousand. With the removal of Newell and Drayer, it ceased to play an active role in engineering-society politics and reform, becoming largely a fraternal organization.[52] The revolt of the younger men had come to an end.

Another engineer who banked on the support of the younger men was Morris L. Cooke. His last major reform in the ASME had been to secure the vote for junior members. In 1923, he sought an electoral comeback in the society when he was nominated to serve on the ASME's governing council. An opposition candidate was put up by the utilities, and Cooke was decisively defeated. Earlier, in 1921, he had been nominated for the presidency of the Philadelphia Engineers Club, but had been defeated by a rival candidate.[53] These two defeats marked the end of Cooke's influence within the engineering profession. He and his friends watched with impotent rage the rising power of business interests, especially the utilities, within the ASME.[54] Significantly, in 1922 President Carman reported that complaints about clique control had ended.[55]

Perhaps in no society was the reaction more complete than in the AIME, which reverted to its pre-1912 position. Almost no articles on the engineer's social obligations appeared in institute publications after 1923. The withdrawal from the FAES was a repudiation of progressive leaders like Hoover and Channing. The reaction against progressivism took the form of a resurgence in the power of the board of directors, and a concomitant decline in membership control of the institute. When it withdrew from the FAES, no ballot of the membership was taken; and the board insisted that it was best able to determine not only the view of the majority but also where the interests of the institute lay.[56] A new constitution was drawn up in 1927 that gave sweeping powers to the board, and it was adopted after only minor modifications. In the first draft proposed, the board could arbitrarily change the institute's bylaws; but, as adopted, the constitution provided only that such changes could be made by the board after a short delay and publication of the proposed modifications. [57]

Alone among the founder societies, the AIEE had no progressive leadership to repudiate. The progressive defeat had come much earlier, in 1912. Leaders like Scott, Pupin, Steinmetz, and others, by the mid-1920s had lost interest in reform. President Berresford's suggestion that policy questions be left to corporation executives met with no opposition; the AIEE lost interest in social responsibility. Prior to 1925, the AIEE had been, at least in form, the most democratic of the founder societies. In that year, however, the open system of nominations was finally abandoned, and the institute adopted a nominating committee. [58] The AIEE entered a period of managed elections and increasing membership apathy. By the 1930s, members were complaining that elections had become a "farce." [59] Perhaps no American engineering society was as overtly controlled by business interests as the AIEE by the later 1920s and early 1930s. As one member put it:

> In the Institute of Electrical Engineers . . . not so long ago one could say, counting on his fingers: "this is a power year, next should be academic and the year after telephone. . . ." Except for a proper sprinkling of university professors and telephone men, committee members and chairmen are responsible intermediate executives of major electrical manufacturers and public utility companies. . . . They represent their companies, I am afraid, more than they do engineers. [60]

The reform spirit among engineers declined less from business opposition than from an internal collapse. Scientific management and conservation, in particular, had provided much of the drive behind engineering progressivism, not only among mechanical and civil engineers, but in the profession generally. They had been important not so much for what they were as what they promised. Before 1920, many engineers thought of them as models for the application of science to human problems and as starting places for the engineering of society. In the 1920s this faith was lost by all but a small minority.

The case of scientific management is perhaps the most distinctive. It was uniquely a movement of engineers; it was assumed by its devotees to be universally applicable to all social processes, and it had aroused widespread hopes among engineers that a new day was about to dawn. Its disintegration as a reform movement in the 1920s marked an end to an era in engineering thought. The basis of engineering hopes had been the claim by Taylor and his followers that they had discovered the laws of management. Of equal importance was Taylor's insistence that his system constituted a single unit which must be accepted or rejected *in toto*. Together these two assumptions led to the belief that engineers would be called upon to reconstruct society. The engineers had the knowledge and only they could apply it properly. But it was precisely these claims which scientific management dropped in the 1920s; it became what Taylor always denied that it was: a collection of efficiency devices.

Ironically, the failure of scientific management as a vehicle of reform coincided with its success in business. Businessmen had never contested the utility of Taylor's work as a set of efficiency techniques; they had resented its claim to absolute managerial power, its arrogant pretensions, and its attacks on traditional management. In 1922, L. P. Alford reported that "the attitude of opposition and mistrust toward management and the passionate antagonism to the installing of management methods have, in general, disappeared."[61] But the hostility disappeared only when manufacturers began to employ college-trained management engineers as a regular part of their staffs. Such bureaucratized engineers did not insist that scientific management was a single system to be adopted as a unit. Management employed Taylor's system piecemeal as a series of efficiency devices; the more ambitious claims of scientific management were quietly dropped by most practitioners.[62] With these changes, most of the old zeal evaporated. Some of the original Taylorites continued the crusade, but they were generals without an army. Their chief accomplishments were several bold systems for labor-management cooperation.[63]

Conservation differed from scientific management in that it was not exclusively a movement of engineers. Conservation retained its vitality as a part of the American reform tradition. But it lost most of its charm for engineers when they discovered that they could not control it. Engineers like Newell had seen in conservation a means of achieving autonomy and social leadership for their profession. By the 1920s, these hopes had largely disappeared. Conservation was as firmly controlled by responsible political officials as was scientific management by business. Although a few engineers were not yet aware of this, notably Arthur E. Morgan, who was to be the first head of the Tennessee Valley Authority, most no longer saw conservation as a means of advancing professionalism. The ASCE, in particular, lapsed into its traditional conservatism. In 1929, the ASCE joined the AEC. But considering the violent opposition of the council to all public enterprise, this step amounted to a repudiation of conservation.[64]

One major consequence of the engineering profession's return to "normalcy" was the separation of reform from professionalism. By the mid-1920s, most engineers hoped to advance professional goals by an alliance with business, rather than by reform. Conversely, those few engineers whose primary loyalty was to reform became divorced from the mainstream of professional development. This was indicated by the later career of Morris L. Cooke. Cooke combined, in a unique way, a commitment to both scientific management and conservation. He linked both of these concerns to American reform politics. But as he gained in stature in reform circles, his influence within the engineering profession declined to zero. By the end of the 1920s, Cooke had become a close ally of Governor Franklin D. Roosevelt. But Cooke no longer looked to the engineering profession for support, and he ceased to link his reform plans to the advancement of engineers.

Cooke's alienation from his profession was indicated by his gradual de-emphasis on the elitist role of the engineer and his renewed stress on democracy in the application of scientific management. By 1919, Cooke had lost faith in the adequacy of the democratic safeguards he had previously supposed to be inherent in scientific management. He sought to democratize the Taylor system by a reinterpretation of the concept of the plan. Cooke now came to argue that it was impossible for the technical expert alone to determine the "One Best Way." This could be found only by the cooperation of all of the interested parties, including labor. For Cooke, the engineer lost his unique role as law giver and became something closer to a mediator. Where Cooke had once drawn a sharp distinction between the "facts" that went into engineering planning and the "opinions" of nonexperts, he now thought that a plan consisted not only of facts, but of judgments based on those facts.[65]

Cooke's democratic reformulation of the original Taylor system reached its culmination in 1928 with his presidential address before the Taylor Society. He maintained that to get real consent and participation by the workers, collective bargaining with unions was essential. His condemnation of yellow-dog contracts and other antiunion devices created a minor scandal within the Taylor Society. But Cooke's revision of the underlying fundamentals of scientific management was even more radical. He tacitly jettisoned not only the "One Best Way," but the conception of absolute, scientific laws of management that was the foundation of the entire Taylor system. Cooke shifted his emphasis from absolute laws to contingent methods. There were no final laws to appeal to, but Cooke insisted that "there must be a general acceptance of the validity of the scientific method from top to bottom."[66]

Cooke's reformation of scientific management led, in his case, to a resolution of the tension between "ought" and "is" that had been so characteristic of engineering progressivism. His abandonment of the idea of absolute laws of management divested technical judgments of their supposed moral significance. Cooke clearly recognized the priority of moral over technical considerations. He warned against "that class of experts who see their problems too narrowly and who place too great dependence upon mathematics in a field where the heart is as much entitled to recognition as is the head."[67] Cooke had once opposed checks and balances in government in the name of efficiency; he now favored them in industry as well as in government in order to serve higher purposes than efficiency. "In the long run," Cooke asked, "if I am not reasonably free, what does it count if I am efficient?"[68]

On the subject of conservation, as with scientific management, Cooke broke away from the narrow confines of professionalism. Cooke's interest in conservation was an outgrowth of his earlier work in the control of public utilities. He now saw cheap electric power as part of a general program of conservation that would advance entire regions. He proposed to junk the many small, privately owned utilities in favor of a few large generating plants located at the coal-mine sites; the resulting cheap electricity would be distributed through an integrated system to a large region. He called this plan "Giant Power."[69]

The technical feasibility of Cooke's "Giant Power" scheme was debatable. Critics pointed out that the coal regions generally lacked sufficient water for cooling the great generating plants Cooke proposed. But there was no question about the change in Cooke's philosophy. Previously he had opposed public ownership, now he favored it. This change was linked to professionalism. Cooke's earlier position was that public ownership was inevitable unless the engineers were put in command. Cooke had seen engineering reform as a way of warding off such socialistic measures as the public ownership of utilities. The "Giant Power" plan of the 1920s was founded on public ownership; it was not a means of elevating the engineering profession generally. Cooke did not expect the engineering profession to support his proposals. Cooke addressed himself to the general public, and he allied himself with liberal politicians like Gifford Pinchot and Franklin D. Roosevelt. Although his "Giant Power" scheme was never adopted, Cooke did have an important impact, notably as first director of the New Deal's Rural Electrification Administration.[70]

In contrast to Cooke, most leaders of the engineering profession reinterpreted the ideology of engineering to give it a conservative meaning.

These formulations stressed the close alliance of engineering and business; and they denied, at least by implication, that the engineer had a significant independent role. Herbert Hoover was no doubt the most prominent spokesman for this point of view in the 1920s. Perhaps second only to Hoover was Michael Pupin. His autobiography was a best seller. In it and in other writings, Pupin presented a version of professionalism that linked the engineer to a conservative, moralistic ideology.[71]

In one sense, Pupin's ideas represented an extreme formulation of the ideology of engineering. Pupin argued that Washington, Franklin, and Jefferson had been engineers and that the early American republic had been guided by the "engineering mind," which he linked to something he called "scientific idealism." Pupin thought this idealism had guided Lincoln and had led to the rise of American universities. He maintained that it was the duty of engineers to preach this gospel and to apply it to politics and theology. But scientific idealism was not a reform philosophy; it was more like a mystical religious insight that would bring man back to God and the nation back to the original principles of Americanism. As Pupin wrote:

> We members of this guild are called upon today to advocate the universal adoption of this idealism. . . . This is the only way leading to a complete demonstration that the spirit of science, engineering, and the industries in league with the Holy Spirit, and not against it. . . . This is a new service which our discipline and training can contribute to the safety of our American democracy.[72]

Pupin's linking of engineering and industry was significant. He drew no sharp line between engineers, scientists, and businessmen. All three were united in a modern-day quest for spiritual enlightenment. Pupin's stress on the moral law—which he identified with mystical Christianity—denied the relevance, if not the existence, of material laws of nature. In an article significantly entitled "The Cosmic Harness of Moving Electricity," Pupin argued that great discoveries came from a search for eternal truth, rather than from material considerations.[73] In Pupin's philosophy, engineers lost their separate identity. By combining science and technology with religious idealism, he reduced the ideology of engineering to vacuity.

A close friend of Pupin's, John J. Carty, the chief engineer of the Bell system, was another important spokesman for engineering in the 1920s. He, too, presented a conservative interpretation of the role and responsibilities of engineers. "The mission of the engineer," Carty maintained, "is to obtain such a mastery in the application of the laws of nature, that man will be liberated, and that the forces of the universe will be employed in his service."[74] But Carty's idea of "liberation" was a rather peculiar one. Future progress was to take place less in the individual than in the group. Society was becoming an "inconceivably great and powerful organism." The chief obstacle to this "inspiring vision" was man himself. Because of man's tendency to "engender conflict and confusion," Carty thought that man threatened further progress. The remedy, Carty maintained, was to accelerate progress by gaining the knowledge "required in preparing the individual man to function as a sane and peaceful unit in the ultimate social organism."[75]

The fear of democracy implicit in Carty's thought was explicit in the publications of other spokesmen for the profession. In 1925, Robert Ridgway, when president of the ASCE, warned of the lingering dangers of paternalism and class consciousness in city government.[76] Similarly, A. A. Potter, a prominent mechanical engineer, noted the lurking threat of "mob control."[77] Nor was Carty alone in suggesting the use of engineering to control men's minds. Francis Lee Stuart, president of the ASCE in 1931, called for a scientific study of man, "the greatest of all raw materials," in order "to make him, including ourselves, ambitious, energetic, helpful, and happy aids in the work of the world."[78] But conservatism in the 1920s was more complacent than fearful. In the confident days of Coolidge and Hoover, social problems no longer appeared to be very pressing or to require any special effort by the engineering profession.

The sanguine temper of engineers in the late 1920s was reflected in a book, *Toward Civilization*, edited by Charles Beard. Beard had edited an earlier volume critical of the machine age, *Whither Mankind;* this was the engineers' reply. All the engineers who contributed were convinced that engineering, science, and business were building a better world. Lee De Forest, for example, thought that improved communications would lead to a better understanding between peoples and ultimately to an end to war.[79] Although most of the writers were willing to admit that technical change created social problems, they argued that the same method could also solve them. Ralph E. Flanders wanted to find a scientific basis for ethics.[80] Dexter S. Kimball brushed aside the complaints of artists and philosophers concerning modern industrial society. He thought that "if in the progress of such a change it be necessary to remodel art, literature, economics, and some of our philosophies of life, I for one am willing to take such a risk."[81]

Not all engineers were complacent in the 1920s. Cooke and a few other progressives were horrified at the alliance between business and engineering. They were deeply concerned over the revival of censorship in engineering societies, the decline in democratic processes, and the dominant position accorded to the spokesmen of business. One of Cooke's friends, William G. Starkweather, wrote to a prominent member of the ASME, C. Harold Berry, protesting the domination of the society by the utilities interests. Berry vehemently denied that the utilities were misusing their power in the ASME. He noted that the utilities had paid his traveling expenses and had provided secretarial and technical assistance to him to support his work on one of the ASME's technical committees. Berry insisted, however, that the business interests in question had not attempted to control his decisions or do anything but contribute to technical progress.[82] There can be little doubt that engineers derived substantial benefits from their alliance with business. But there was a danger that in gaining worldly things the engineering profession might have lost its own soul.

NOTES

1. Committee on Elimination of Waste in Industry, *Waste in Industry* (New York, 1921), v–vii, 4. The internal organization of the committee may be followed in boxes 1-I/582, 583, 584, and 585, Herbert Hoover Papers, Herbert Hoover Presidential Library, West Branch, Iowa.

2. L. W. Wallace, "Industrial Waste," *Journal AIEE*, XL (November, 1921), 868–871.

3. Edward E. Hunt to L. W. Wallace, December 5, 1921, file "Unemployment, FAES," box 1-I/410, Hoover Papers.

4. Edward E. Hunt to Morris L. Cooke, September 4, 1922, file "Unemployment, M. L. Cooke," box 1-I/410, Hoover Papers.

5. Committee on Elimination of Waste, *Waste in Industry*, 3–8, *passim*.

6. *Ibid.*, 9–10.

7. *Ibid.*, 3, 8.

8. *New York Times*, June 4, 1921, p. 7. See also "Digest of June 3 Board Meeting, The Federated American Engineering Societies" (mimeograph), pp. 7–8, file "Federated American Engineering Societies," box 1-I/128, and Edward E. Hunt to L. W. Wallace, June 22, 1922, and July 3, 1922, file "Federated American Engineering Societies," box 1-I/410, Hoover Papers.

9. "Herbert Hoover Emphasizes Duty of Engineers in Reestablishing Economic Balance," *Mechanical Engineering*, XLIV (March, 1922), 206–207.

10. Morris L. Cooke, *Professional Ethics and Social Change*, (New York, 1946), 15.

11. The Thompson & Lichtner Company to L. W. Wallace, March 14, 1922, and M. W. Alexander to E. J. Mehren, February 3, 1922, file "Federated American Engineering Societies," box 1–I/410, Hoover Papers.

12. Edwin Ludlow, "Adjustment of Wages and Working Conditions," *Mining and Metallurgy*, II (March, 1921), 3.

13. Bradley Stoughton to Calvert Townley, June 29, 1921, file "A.I.M.M.E., 1921–1922," box 1–I/20, Hoover Papers.

14. See, for example, W. R. Ingalls, "Waste in Industry," *Mining and Metallurgy*, III (January, 1922), 1–2.

15. T. A. Rickard, "Institute Affairs," *Engineering and Mining Journal*, CXV (May 26, 1923), 922.

16. "Excerpts from Address of President Berresford at Opening Session," *Journal AIEE*, XL (March, 1921), 251.

17. Arthur W. Berresford, "Personal Observations in the Industry," *Journal AIEE*, XL (July, 1921), 555.

18. *Ibid.*, 555, 557, 560.

19. Kenneth E. Trombley, *The Life and Times of a Happy Liberal, A Biography of Morris Llewellyn Cooke* (New York, 1954), 97–100. See also the material in file "Drury-Steel Investigation," box 179, Papers of Morris L. Cooke, Franklin D. Roosevelt Library, Hyde Park, New York, and Committee on Work Periods in Continuous Industry, *The Twelve-Hour Shift in Industry* (New York, 1922), 3–4.

20. "F.A.E.S. Report States Facts on Twelve-Hour Shift," *Mechanical Engineering*, XLIV (October, 1922), 681.

21. Hoover to Harding, November 1, 1922, file "Twelve-Hour Day," box 1–I/313, Hoover Papers.

22. Committee on Work Periods in Continuous Industry, *The Twelve-Hour Shift in Industry, passim*.

23. Commission of Inquiry, The Interchurch World Movement, *Report on the Steel Strike of 1919* (New York, 1920), had served to arouse public opinion on the issue.

24. Edward E. Hunt to L. W. Wallace, June 22, 1922, file "Federated American Engineering Societies," box 1-I/410, Hoover Papers.

25. "The Civil Engineers and Society Co-Operation," *Engineering News-Record*, XC (April 12, 1923), 652–653.

26. W. R. Ingalls, "The Twelve-Hour Shift in Industry," *Mining and Metallurgy*, IV (March, 1923), 115, and "Elimination of the Twelve-Hour Day in the Steel Industry," *Mining and Metallurgy*, IV (June, 1923), 304–305.

27. Channing to L. W. Wallace, November 17, 1921; Hoover to Channing, November 22, 1921; and Channing to Hoover, November 25, 1921, file "J. Parke Channing," box 1-I/14, Hoover Papers.

28. F. F. Sharpless to representatives of AIME on American Engineering Council, January 12, 1924, file "A.I.M.M.E. 1923-1924," and "Minutes of Meeting, AIME Board of Direction," October 20, 1922 (mimeograph copy), p. 2, file "A.I.M.M.E., 1921-1922," box 1-I/20, Hoover Papers.

29. F. F. Sharpless to representatives of AIME on American Engineering Council, June 2, 1923, file "A.I.M.M.E. 1923-1924," box 1-I/20, Hoover Papers. See also "The Institute and the Federated," *Mining and Metallurgy*, V (August, 1924), 399–400.

30. Channing to Board of Directors, AIME, January 25, 1924, file "J. Parke Channing," box 1-J/14, Hoover Papers.

31. Channing to Hoover, June 2, 1923, file "J. Parke Channing," box 1-J/14, Hoover Papers.

32. "Local Sections and Affiliations," *Mining and Metallurgy*, V (August, 1924), 406.

33. *Ibid.*, 404–407.

34. M. E. Cooley to L. W. Wallace, August 20, 1922, file "Federated American Engineering Societies," box 1-I/410, Hoover Papers.

35. M. L. Cooke, *Professional Ethics and Social Change*, 15.

36. "Gardner S. Williams Reviews Last Revision of New Constitution," *American Engineering Council Bulletin*, III (April, 1924), 5. See also "F.A.E.S. Changes Name to Engineering Council," *Engineering News-Record*, XCII (January 17, 1924), 129.

37. Edward E. Hunt to L. P. Alford, September 10, 1921, file "Federated American Engineering Societies," box 1-I/128, Hoover Papers.

38. J. C. Ralston, "Our Industrial Ideal," *Bulletin Federated American Engineering Societies*, II (September, 1922), 5-6.

39. "Federated American Engineering Societies," *Engineering Education*, XIII (May, 1923), 584-585.

40. "Prospectus of Five-Year Program of Research by American Engineering Council" (typescript), file "American Engineering Council," box 1-I/19, Hoover Papers. See also American Engineering Council, Scope and Activities, 1924-1940, "Character and Scope of Activities of American Engineering Council," April 21, 1926, p. 5, American Engineering Council Papers, Engineering Societies Library, New York (hereafter cited as AEC Papers).

41. *American Engineering Council Bulletin*, V (January, 1926), 4.

42. "American Engineering Council," *Journal AIEE*, XLVII (February, 1928), 157.

43. *Ibid.*

44. Herbert Hoover, *The Memoirs of Herbert Hoover*, 2 vols. (New York, 1952), II, 61-62.

45. American Engineering Council, Executive Committee Minutes, I, June 3, 1926, pp. 2-3, AEC Papers.

46. American Engineering Council, Administrative Board Minutes, II, January 15, 1925, p. 9, AEC Papers.

47. American Engineering Council, Executive Committee Minutes, I, April 1, 1926, p. 6, AEC Papers.

48. American Engineering Council, Administrative Board Minutes, II, October 19, 1928, p. 8, AEC Papers.

49. American Engineering Council, Executive Committee Minutes, II, May 15, 1931, Appendix A, pp. 1-6, and January 14, 1932, pp. 18-19, AEC Papers.

50. *Ibid.*, January 12, 1933, Appendix, pp. 15-16, AEC Papers.

51. "A.A.E. Rejects Plan for Salaried President," *Engineering News-Record*, LXXXVI (May 19, 1921), 870-871, and "American Association of Engineers Convention Passes Stabilizing Measures," *Engineering News-Record*, LXXXVIII (June 15, 1922), 1011-12. For Drayer's own version of what happened, see Drayer to Morris L. Cooke, November 12, 1927, file 562, box 49, Cooke Papers.

52. "A Turning Point," *Engineering News-Record*, XC (May 17, 1923), 859, and Frederick Haynes Newell, "Decentralization and the Association's Experience," *Engineering News-Record*, XC (June 7, 1923), 1014.

53. Cooke to Guido H. Marx, August 1, 1923, file 263, box 23, and William G. Starkweather to H. K. Hathaway, October 8, 1923, file "ASME Material," box 169; C. V. Lathrop to Cooke, May 21, 1921, and Henry Hess to John C. Trautwine, Jr., May 5, 1921, file 332, box 28, Cooke Papers.

54. William G. Starkweather to Cooke, December 17, 1931, and November 23, 1932, file 377, box 85, Cooke Papers.

55. "President Carman's Address," *Mechanical Engineering*, XLIV (January, 1922), 8, 74.

56. "The Institute and the Federated," *Mining and Metallurgy*, V (August, 1924), 401.

57. "Tentative Draft of Revised Constitution and By-Laws," *Mining and Metallurgy*, VIII (September, 1927), 374–380; "Vote for the Constitution," *Mining and Metallurgy*, VIII (December, 1927), 494.

58. "Revision of A.I.E.E. Constitution," *Journal AIEE*, XLIV (January, 1925), 87–88.

59. B. M. Jones, "Membership Participation in AIEE Election" (letter to the editors), *Electrical Engineering*, LIX (April, 1940), 168.

60. John Mills, *The Engineer in Society* (New York, 1946), 136.

61. L. P. Alford, "Ten Years' Progress in Management," *Mechanical Engineering*, XLIV (November, 1922), 700.

62. "Progress in Management Engineering," *Mechanical Engineering*, XLVIII (December, 1926), 1407–09, and Dexter S. Kimball, "Has Taylorism Survived," *Mechanical Engineering*, XLIX (June, 1927), 593–594.

63. Milton J. Nadworny, *Scientific Management and the Unions, 1900–1932* (Cambridge, 1955), 122–141.

64. "ASCE Joins Engineering Council," *American Engineering Council Bulletin*, VIII (May, 1929), 8. The history of conservation may be followed in Samuel P. Hays, *Conservation and the Gospel of Efficiency* (Cambridge, 1959), and Donald C. Swain, *Federal Conservation Policy, 1921–1933* (Berkeley and Los Angeles, 1963).

65. Morris L. Cooke, *Our Cities Awake* (New York, 1919), 97–98.

66. Morris L. Cooke, "Some Observations on Workers' Organizations," *Bulletin of the Taylor Society*, XIV (February, 1929), 8.

67. *Ibid.*

68. *Ibid.* Cooke was by no means unique in his revisionism. For the evolution of scientific management thought, see Nadworny, *Scientific Management*, 97–121.

69. Morris L. Cooke, "Giant Power: An Interpretation," *The Atlantic Monthly*, CXXXVIII (December, 1926), 813–822.

70. For Cooke's disgust with his own profession on the power issue, see [Morris L. Cooke], "The Price System and the Engineer," *The New Republic*, LIII (January 25, 1928), 261–262. Cooke's anonymous papers are to be found along with his signed ones in the seven bound volumes of his publications in the Cooke Papers. For Cooke's later career and his relationship to Roosevelt, see Trombley, *Life and Times*, 108–159, 164, *passim*.

71. Michael Pupin, *From Immigrant to Inventor* (New York, 1930); Michael Pupin, *The New Reformation, From Physical to Spiritual Realities* (New York, 1927); and Michael Pupin, *Romance of the Machine* (New York, 1930).

72. Michael Pupin, "Engineering and Public Service," *Journal AIEE*, XLIV (October, 1925), 1044.

73. Michael Pupin, "The Cosmic Harness of Moving Electricity," *Journal AIEE*, XLV (August, 1926), 758–760. See also Pupin, *New Reformation*, 181, and Michael Pupin, "The Power Age and Modern Civilization," *Electrical Engineering*, LI (March, 1932), 156–157.

74. John J. Carty, "Ideals of the Engineer," *Journal AIEE*, XLVII (March, 1928), 211.

75. *Ibid.*, 211, 212.

76. Robert Ridgway, "The Modern City and the Engineer's Relation to It," *Trans ASCE*, LXXXVIII (1925), 1251–53.

77. A. A. Potter, "The Engineering College—Its Opportunity for Service," *Journal of Engineering Education*, XVI (September, 1925), 17.

78. Francis Lee Stuart, "The Engineer's Growing Civic Responsibilities," *Trans ASCE*, XCV (1931), 1301.

79. Lee De Forest, "Communication," in Charles A. Beard, ed., *Toward Civilization* (New York, 1930), 136.

80. Ralph E. Flanders, "The New Age and the New Man," *ibid.*, 33.

81. Dexter S. Kimball, "Modern Industry and Management," *ibid.*, 139.

82. C. Harold Berry to William G. Starkweather, June 16, 1933, file 377, box 85, Cooke Papers.

10 ▪ DEPRESSION
AND NEW DEAL:

THE ENGINEERS' IDEOLOGY IN DECLINE

The Great Depression was at once the engineers' greatest challenge and greatest opportunity. Engineers had long claimed that they had both the knowledge and the responsibility for solving the problems created by modern technology, and the depression clearly was such a problem. Public opinion was ripe for an attempt to engineer society. This was indicated by the widespread interest in technocracy. It was reflected on a more responsible level by sober proposals for comprehensive national planning by leaders of the business community, such as Gerard Swope, the president of General Electric, and the United States Chamber of Commerce. It found expression in the planning experiments of the early New Deal.[1] Had the engineers been able and willing to come before the public with a clear-cut blueprint for restoring prosperity, they would have received a hearing. They might have assumed a large role in national leadership. But the profession neither responded to the challenge nor did it take advantage of the opportunity.

Engineers, ironically, played only a very minor direct role in the planning activities of the 1930s. The reforming engineers of the previous generation would have regarded the new shape of public opinion as a heaven-sent opportunity, and they almost certainly would have acted. But by 1929, the profession was dominated by business, it lacked effective leaders for such progressive sentiment as existed, and it was deeply divided. Perhaps more importantly, engineers were losing interest in social reform. As the 1930s wore on, engineers turned away from social responsibility as a means of advancing professionalism. While the public clamoured for leadership, the engineers were bankrupt of ideas, incapable of action, and obsessed with their own immediate selfish interest.

If the direct impact of engineers on planning was small, their indirect influence was much larger. Engineers were an important source of the planning ideas current in America in the early 1930s. The degree of their influence, however, is not easy to evaluate precisely. This is particularly the case with liberal reformers. It is possible to show that the engineers' ideas reached several important intellectuals, including Thorstein Veblen, Charles Beard, and others.[2] Several New Deal leaders were influenced by the engineers, notably Henry Wallace.[3] It is the degree of influence that is doubtful. Liberals and engineers derived ideas of planning and social reconstruction from similar sources; both, for example, were deeply influenced by evolutionary ideas. Planning was inherent in pragmatism, in socialism, and in other modern philosophies. The planning experience of America and other nations during the First World War was a further source of inspiration. Engineers had an influence in the development of liberal thinking on the subject of planning, but it is doubtful whether the results would have been significantly different had the engineers confined themselves exclusively to their technical affairs.

The case is somewhat clearer for the business community. Engineers played a very important role in the slow transmission of planning ideas to business. Herbert Hoover, more than any other man, educated businessmen to a larger social philosophy. His activities as Secretary of Commerce represented a large-scale attempt to put into action some of the fundamental ideas of engineering thought, such as planning, efficiency, and rationalization. Other engineers in high positions, like Gerard Swope, were also important carriers of these ideas. Businessmen were, of course, strongly influenced by experience with wartime planning as well as by trade-association activities.

With the coming of the depression, businessmen in great numbers turned to national planning for a solution to the nation's problems. The extent to which business had been won over was suggested by a questionnaire administered by a group of engineering educators in 1931. About one hundred top business leaders were surveyed; sixty-five thought that the peacetime economy should be so systematized as to supply the needs of the people at all times. Only seven were against this sort of planning. The great majority felt that this end could be achieved without government ownership through voluntaristic activities by business itself. Although other sources of planning ideas existed, there can be little doubt that in this case the engineers were a major source of inspiration.[4]

The technocracy fad in 1932 illustrated the indirect influence of the engineers. On the face of it, technocracy was an engineers' movement. Its leader, Howard Scott, claimed to be an engineer; its philosophy, in essence, was to end the depression by a dictatorship of the engineers. In practice, though the ideas were derived in large part from the engineering progressivism of the First World War, the movement had only a few engineers among its leaders, and most of these severed their connections with it within a few months. It represented a grotesque parody of the engineers' thought, rather than a legitimate expression of it.

The origins of the technocracy movement go back to 1919, and specifically to Thorstein Veblen's activities at the New School. Veblen's interest in engineers had been aroused by his rather limited knowledge of the ferment within the ASME. He knew something of Henry L. Gantt's philosophy, and that Gantt had formed a discussion group, the "New Machine," which had been left leaderless by his death. He had read some of Morris L. Cooke's writings, and paraphrased them in his *Engineers and the Price System*. The activities of Gantt and Cooke apparently convinced Veblen that the engineers had a revolutionary potential which the American labor movement seemed to lack. Becoming "obsessed" with engineers, Veblen persuaded a friend of his, Guido Marx, a mechanical engineer from Stanford, to come to New York and attempt to organize the engineers.[5]

Veblen and his friends never succeeded in gaining any real support from the engineers. Cooke cooperated by putting Marx in contact with a number of engineers, and Cooke had several discussions with Veblen. The topic apparently was Cooke's ideas on electric power, which he later presented in his Giant Power scheme. Cooke regarded Veblen as a spokesman for the "extreme left" and showed little interest in being converted to Veblen's way of thought. A group did gather at the New School, however, and it included a few engineers who had been active in Gantt's New Machine. The focus of its activities was the course taught by Marx and, after his return to Stanford, by Otto S. Beyer, another left-wing engineer. This group, the Technical Alliance, included people who were not engineers, notably Howard Scott. It was out of the discussions of this group that the ideas of technocracy were born.[6]

The popular craze for technocracy was due in part to Walter Rautenstrauch, a former Gantt follower who had been active in the New Machine. He was a member of the Department of Industrial Engineering at Columbia University, and he was able to provide some rooms in that department for an "energy survey of North America," which some of Scott's followers had organized. The news that a group of engineers at

Columbia had the answer to the depression aroused widespread popular interest and made "technocracy" a household word for a short period.[7]

The technocracy bubble burst almost as quickly as it had appeared. Howard Scott was shown to be not an engineer but a denizen of Greenwich Village. Scott's controversial personality and utterances soon split the movement into fragments. A motley group of left wingers organized the Continental Committee on Technocracy, but it lacked organizational coherence and soon disintegrated. Scott and his followers built a close-knit, disciplined organization, Technocracy Incorporated. Its use of grey uniforms, monad symbols, and its anticatholic bigotry aroused apprehensions on the part of some that technocracy was a native American totalitarianism.[8]

After the first few months, reputable engineers played almost no role in the technocratic agitation. Rautenstrauch and Jones were engineers in the old progressive tradition. They broke with the movement and attempted to defend their professional reputations. Both insisted that what they favored was public control, not public ownership, of industry.[9] Rautenstrauch for a time attempted to lend respectability to the technocratic idea; he argued that its true aim was "to suggest a design of society in which class struggle is impossible."[10] These efforts were not successful. The movement was roundly condemned by the American Engineering Council.[11]

For all of its zaniness, the technocracy movement did show the large reservoir of public interest that was there for the engineers to tap. A responsible attempt by the engineering profession to cope with the depression surely would have aroused interest. That no such effort was made must be accounted one of the profession's lost opportunities. One reason for this failure lay in the technocracy movement itself. It tended to discredit the profession's reform tradition. Anyone who proposed to engineer society opened himself to ridicule by his colleagues.[12]

Another reason for the failure of the engineers to capitalize on the depression was that they were as bewildered as most of the rest of the population. They had not anticipated the collapse, and they had no plans for dealing with it. Engineers found it hard to accept the fact of the economic debacle. Until 1931, engineers, as indicated by the technical press, ignored or tried to minimize the crisis. When it could no longer be ignored, the first reaction was defensive. H. P. Charlesworth, president of the AIEE, argued in 1933 that America really had experienced a century of progress and that science had led to an increase in human happiness in the broadest sense.[13] A former president of the AIEE maintained the depression was a temporary dislocation that would pass as had other depressions in history.[14]

But the evidences of waste, misery, and discontent could not simply be explained away. In an unprecedented action, the presidents of all four of the founder societies wrote a joint introduction to a series of articles on "Engineering and Human Happiness," sponsored by the Engineering Foundation. In this introduction, though they painted a glowing picture of the many tangible benefits engineering had conferred on humanity, they did cautiously suggest that "perhaps the methods of engineering and science could be applied beneficially to some features of this problem also, supplementing progressive business methods."[15] But the first article in the series, by C. Kenneth Mees, was somewhat more pessimistic. He thought that engineering had not led to human happiness.[16]

Engineers in America had for so long been taking credit for progress and prosperity that their first reaction was a sense of guilt. A prominent mechanical engineer, C. F. Hirschfeld, thought that "this cloud is of our own making," and he maintained that engineers and scientists had "almost criminally refused to give serious thought to the collateral results" of their work.[17] Only slowly did engineers begin to realize that the crisis might be an opportunity for their profession. Roy V. Wright, as president of the ASME, welcomed outside criticism of engineers, since it would stimulate a sense of social responsibility on the part of engineers. He urged engineers to take the lead in solving the nation's difficulties.[18] Similarly, Frederick M. Becket of the AIME in 1934 urged engineers to exercise the same sort of stewardship over their knowledge that the philanthropist demonstrated in the use of his money.[19] But these pleas were vague and unspecific.

Another reason for the failure of the engineers to provide leadership in the depression was the lack of consensus within the profession. A few voices within the profession called, rather timidly, for action. But those who favored action and leadership had to face the fact that, as a result of the conservative reaction in the 1920s, the profession was firmly controlled by business-oriented engineers. These men either favored no action at all, or preferred to let the business community do the acting. The spokesmen for the various private utilities were particularly powerful in all the major engineering societies. Probably because their industries were vulnerable to demands for public ownership or drastic public control, such men tended to stand for *laissez faire*. One utilities executive, William S. Lee, in 1931 stated that the engineer's duty was to maintain initiative, individualism, and the type of citizenship "that recognizes fully the fact that the only true equality in life is equality in opportunity."[20] Specifically, Lee thought that the engineer should educate the public in order to prevent the govenment from "entering business in competition with its citizens and thus embark upon a course disastrous to the best interests both of those directly affected and of the taxpayers of the country in general."[21] Another utilities spokesman, Arthur W. Berresford of the AIEE, maintained that planning was not a possible solution to the depression and that the only hope lay in the efforts of millions of separate individuals guided by intelligent self-interest.[22]

The paralysis of the engineering profession was exemplified by the inaction of the American Engineering Council. In the early 1930s, there was some talk in engineering circles of using the council in order to bring the engineering viewpoint to bear on the problems of the nation; this had in fact been the purpose in founding the council. But the conservative counterrevolution, which had converted the Federated American Engineering Societies into the American Engineering Council, had made it all but impossible for the council to take any positive action not approved by the ultraconservative minority. When the ASCE joined the council in 1928, an effort was made to relax somewhat the veto rights of council members. Before any important action could be taken, the assent of one-half of the delegates of each of the founder societies was necessary; previously the figure had been two-thirds. Similarly, all policy decisions now had to have the support of one-half rather than two-thirds of the aggregate of the smaller national societies and of the total of regional and local societies.[23] But this liberalization of the constitution appears to have had no practical effect. The AEC continued to take a position more conservative than that of the business community and unrepresentative of the engineering profession.

In the early 1930s, when many business leaders were calling for national planning and other departures from orthodoxy, the AEC adhered to strict *laissez faire*. The council took the extreme position that the Air Corps should not be allowed to test its own equipment because this would let a government agency compete with private enterprise. The council treated *laissez faire* as an absolute principle that admitted no exceptions. It opposed a bill to regulate the wages of construction workers as a violation of economic law.[24] The council was in no position to take the lead in the national debate over how to end the depression.

The council's only important activity directed at seeking a cause for the depression was a study, authorized in January, 1931, of the relations of consumption, production, and distribution. Three of the five men who made this study were progressives associated with the scientific-management movement: Ralph E. Flanders, Leon P. Alford, and Dexter S. Kimball. This committee's first report, presented in February, 1932, favored "taking such chances as may be involved in developing a rationally controlled economy." With many qualifications, the committee advocated countercyclical financing; it opposed detailed governmental control and approved the development of the trade associations as instruments of planning. The committee strongly endorsed the necessity and desirability of the profit motive, admitting of its restraint only in extreme cases where it had become "destructive and unsocial."[25]

The committee's study was anything but a clarion call to the profession. It did not suggest an engineering solution to the depression; nor did the committee call upon the profession to assume political leadership. In substance, the report was a pale shadow of the proposals being discussed within the business community. It was, for example, much less daring or imaginative than the plans of Gerard Swope and the Chamber of Commerce. The council regarded the committee's report as highly "debatable," and it took no action to implement its various recommendations.[26] Although unable to formulate its own program, the AEC rejected that of Herbert Hoover. It opposed his proposals to stimulate the construction of public works by loans from the Reconstruction Finance Corporation.[27]

The founder societies were considerably more flexible than the American Engineering Council. All except the AIME opened their publications to papers on social and economic topics. But the talk did not lead to significant action. The ASME was virtually the only society to take any official action at all. Due mainly to the promptings of Roy V. Wright, an "Engineers Civic Responsibility Committee" was formed. Almost its only accomplishment was to prepare a manual on citizenship for distribution to junior members.[28] Unlike the situation in the period 1908–1918, there were no organized opposition parties formed, and no threat developed to the existing rulership of the societies.

The shift left of the New Deal in 1935 led to a halt in the publication of unorthodox papers. Censorship was restored. The AIEE, for example, had printed many papers on the depression and on professional questions in the early 1930s. In 1935, a committee of the institute, headed by Arthur W. Berresford, urged a tighter control of publications. The committee recommended that all papers be "subject to scrutiny and acceptance by committees organized to pass upon the value, desirability, and pertinence of the matters contained." The committee affirmed the institute's traditional policy of prohibiting "speculative" papers and "abstract propositions" dealing with such subjects as rate making and utility regulation. [29]

In 1936, a number of members of the AIME proposed the organization of a special division for the discussion of social and economic subjects. "If such a Division were in existence," one member commented, "the relentless, persistent, and effective hostility of the Board could be confidently predicted." [30] Even the ASME, traditionally the most liberal of the founder societies in its publications policies, felt the pressure. Marion B. Richardson, who did a great deal of committee work relating to program activities of the ASME, noted in 1942, that "there has always been a feeling among many of us, especially with papers dealing with economic and professional status subjects, that the only papers that 'could get by' were those that presented 'facts' we liked to hear." [31]

Three men illustrated the transformation of the attitudes of engineers toward the depression after 1935. Ralph E. Flanders, William E. Wickenden, and Arthur E. Morgan represented, respectively, mechanical, electrical, and civil engineering. They were original and articulate. Flanders was a businessman, the head of a New England machine-tool firm. Wickenden had been a telephone executive before becoming the president of Case Institute of Technology. Morgan headed a civil-engineering firm before accepting the presidency of Antioch College. Together these three were the most important spokesmen for the progressive tradition in engineering during the early 1930s. But unlike Cooke, Gantt, and Newell, these men were not rebels. In the later 1930s, they lost their faith in not only the New Deal, but in the idea of an engineering solution to social problems as well.

One of the ablest proponents of planning in the early 1930s was Ralph E. Flanders. He had been deeply influenced by scientific management, and he thought science might provide the basis for solving social problems. As early as 1930, Flanders urged that the "time has come, then, to apply the engineering viewpoint to the social machine." [32] In 1931 he argued that "the same investigation of and adjustment to natural laws, and the same critical analysis of the conditions of the problem which have served the engineer in the physical field, should produce results of comparable effectiveness in the social environment." To carry out this idea he proposed the creation of a new professional type, the "social engineer." [33] Flanders believed that "our safety and progress from this time forth depend on self-conscious, carefully planned, purposeful action." [34] As late as September, 1934, Flan-

ders spoke admiringly of President Roosevelt, thinking that he was "devoted to the task of fruitful social reconstruction."[35]

With the New Deal's shift left in 1935, Flanders's view began to
change drastically. He became openly critical of the New Deal, maintaining that because reform was "directed against profits, our mass of
unemployment must continue."[36] The Tennessee Valley Authority he
saw not as a "yardstick" but as a "knotted club" used against business.[37] By 1936, Flanders had lost his faith in a "scientific" solution to
the depression. He thought that correct moral principles were the foundation for all progress, and he drew up a list of precepts for each class in
the community.[38] Flanders showed an increasing concern for the traditional principles of American individualism. He thought that an equivalent to the frontier must be discovered in order to "make it possible for
us to retain our constructive national psychology."[39] In 1937, he repudiated the idea of national planning.[40] In 1940, he denounced those who
were "rendering lip service to capitalism while installing continuously
one after another of the elements of state socialism."[41]

Like Flanders, William E. Wickenden was appalled by the depression. He spoke eloquently of "the sad dilemma of a hungry world in
which farmers burn the crops for fuel, of ragged men in the shadow of
mills . . . of a business world as much troubled with unemployed dollars as with unemployed men."[42] He concluded that "rugged individualism is giving way before a broad socializing movement."[43] He took a
pragmatic view of capitalism; it had been based on the assumption of an
expanding economy, but since this condition no longer held, major
changes were necessary.[44]

As an engineering educator, Wickenden was not concerned with the
details of national politics. The focus of his concern was the intellectual
fundamentals of engineering: the social role and social responsibilities of
the engineer. But he was as much influenced as Flanders by the changing attitude of the New Deal to business. By 1938, he believed that
"behind all other social issues looms the overshadowing struggle, now
worldwide, between collective security and free enterprise."[45] Wickenden thought that New Deal policies were halting the flow of capital to
industry, and he urged the engineer to "reckon your stake in free institutions and in free enterprise."[46]

Wickenden saw the engineer as the guardian of individualism. He
thought that the key to engineering success lay in enterprise, which he
defined as "the inner drive which urges men to get on and not merely
to hold on, to depend on their own efforts."[47] He advised young engi-

neers that "if you succeed it will be because you are spurred by inner drives rather than outside rewards."[48] To Wickenden, the essence of professionalism was its practitioners' willingness to make a supreme effort, to go a "second mile."[49] This idea left engineers with no separate identity, scientific or otherwise, and grouped them with the mass of ambitious organization men in business.

In many ways the most interesting of the spokesmen for engineering in the 1930s was Arthur E. Morgan. His intellectual Odyssey demonstrated the bankruptcy of engineering progressivism. Morgan, the first chairman of the Tennessee Valley Authority, had long been a strong advocate of planning. He defended the T.V.A. against the charge of socialism. He maintained that the American philosophy of life was pragmatic and that theory should be modified to fit experience. Thus, he believed that private initiative had given way where less useful than public enterprise, as in the case of schools, fire departments, and roads. He concluded that "in view of this vast and ever-increasing experience of the public in business for itself, it is difficult to persuade the average American that there is something sinister and evil about public business and something divinely inspired and holy about private initiative." Morgan was convinced that "the difference between civilization and savagery is social and economic planning."[50]

In 1938, President Roosevelt removed Morgan from his position as head of the T.V.A. Several matters of substance were involved in Morgan's ouster. Morgan conceived the T.V.A. as a sociological laboratory; others, notably David Lilienthal, were more concerned with economic growth than with social engineering.[51] Morgan was unhappy with the overall drift of the New Deal, especially what he regarded as its antibusiness bias and its tendency to undermine traditional values. These issues found expression in disputes between Morgan and Lilienthal on the proper relationship of the T.V.A. to the private utilities. Morgan's position, which he made clear after his departure from government service, was that paternalism threatened American character. He urged that the public should pay the full cost of such services as housing, water, and electricity, since "otherwise where shall we stop?"[52] A further source of difficulty was Morgan's insistence on professional autonomy for engineering work.[53]

Although Morgan broke with the New Deal, he did not abandon the dream of an engineered society. He saw in planning a means of preserving traditional values. Like other engineers, Morgan feared that "the flux of modern life" was destroying these "common values."[54] But unlike other engineers, Morgan's proposals for preserving them were not simply an affirmation of moralism. He saw the small town as the institutional carrier of these values; he proposed that such towns be preserved and defended through a system of regional planning. These regional groupings would be "a key to a new social and economic order."[55]

But Morgan's hopes of regional planning to preserve the small town and its values ran into conflict with the idea of democracy. Morgan's dilemma was an old one for engineers, between planning by experts and the need for popular control in a democratic society. Morgan faced the issue squarely. Democratic control could not be accepted. His own experiences with the T.V.A. had shown that in this way planning might be diverted from what he considered its proper aim. Democracy would lead to competition between sovereignties and ultimately to the violation of the ideal of "natural" regions. Morgan admitted that such an organization of society might lead to totalitarianism, but he hoped that this could be avoided by strictly observing moral principles and using coercion "only when other means of co-ordination fail, and when the general public interest cannot otherwise be protected."[56]

Morgan's vision of an authoritarian regionalism failed to arouse public interest. He soon abandoned it. Morgan, however, continued his crusade for the small town. He did this by his writings as well as by his private consulting practice. Morgan never lost faith in social engineering, as shown by his admiration for the utopianism of Edward Bellamy. But in his later writings, Morgan demonstrated an awareness of the dangers of despotism and an apprehension of the misuses of power.[57] In any case, Morgan was so absorbed in the small town that he lost sight of that other type of community, the engineering profession.

Flanders, Wickenden, and Morgan were the most conspicuous examples of a general retreat from planning by the engineering profession. The converse of this reaction against progressivism was a return to the idea that it was the engineers' responsibility to defend free enterprise and traditional beliefs. One president of the ASME lamented the "decay and overthrow of institutions." He thought it necessary for "thinking people" like the engineers to lead the nation back to "the faith of our fathers."[58] An ASCE president, Malcolm Pirme, in 1944, drew up a statement of principles which he hoped would constitute a "chart for the future," leading to the reestablishment of free enterprise.[59] The return to faith and the codification of moral principles became the order of the day.

The reaction against planning was lasting. It was reinforced after the Second World War by the general conservative tendency of public opinion. The surprising success of Friedrich von Hayek's *The Road to Serfdom* was one indication of the drift in social thought; the popularity of George Orwell's nightmarish vision of a totalitarian anti-Utopia constituted another. The cold war served to identify planning with totalitarianism and communism.[60] After 1945, engineering opinion faithfully reflected these trends.

Despite the conservative reaction, the underlying ideas of the engineers' ideology showed a surprising vitality, at least verbally. Engineers continued to identify their profession with scientific and technological progress, they portrayed themselves as impartial middle men, and they stressed the idea that engineers bore a social responsibility for solving modern problems. But there was an increasing element of doubt in their assertions. By the 1940s, these themes had been reduced to little more than a ritual. They made engineers feel good. They indicated the continuing strength of professional values among engineers. But they led to no action; social responsibility was moribund.

With the decline of the idea of social responsibility, engineers found new ways of expressing their professional aspirations. The real battles within the engineering profession did not center on the reconstruction of society, but on the bread-and-butter issues of how to advance the specific interests of engineers as a group. As earlier, there was a lack of consensus on how to proceed. Engineers might look for models to the traditional professions, to organized labor, to business, or to science. As a crude first approximation, it might be said that the dominant professional strategy was alliance with business. This was challenged in the 1930s and 1940s by licensing. The leading issue in the 1950s was collective bargaining. Scientific professionalism has come into prominence since 1960. A full examination of the development and competition of these ideas would carry us far beyond 1940 and lies outside the scope of this study. But a brief sketch, even though necessarily inadequate, will help explain developments in the 1930s and shed light on the decline of the idea of social responsibility and the growing divorce of professionalism from social reform.

Licensing and other devices for "closing" the engineering profession gained many adherents among engineers in the 1930s. The idea was, of course, far from new. This was the means used by doctors and lawyers to protect their professions. The idea had been agitated among engineers since the days of the Technical League. Licensing had provided much of the dynamism of the American Association of Engineers in its period of great influence. Although interest in licensing declined during the prosperous 1920s, it was not forgotten. The depression was a particularly favorable time for a revival. Clearly one of the major problems of the 1930s was an oversupply of engineers; too many engineers were chasing too few jobs. Licensing, which aimed at limiting the supply of engineers, seemed to offer a remedy to some of the profession's worst problems.

As in earlier battles over licensing, the support came most heavily from discontented younger men, and the leadership was provided by independent consultants in civil engineering. The unhappiness of the younger men was easy to understand. In the period between 1929 and 1933, the income of engineers declined almost twice as much as the average of all salaries.[61] It was the younger men who were hardest hit by unemployment and low salaries. The consultants in civil engineering felt especially threatened in the 1930s, because so much of the remaining work in their field was undertaken by the federal government. The private consultant appeared to be facing extinction.[62] David B. Steinman, a very distinguished designer of bridges, emerged as the leading spokesman for closing the engineering profession by means of licensing.

Licensing agitation met firm opposition from the leaders of the founder societies. Many engineers considered licensing a form of collectivism little different from unionism. One engineer wrote in 1922, that "most of us have never been willing to stand for labor-union principles and this seems to be the adoption of their foremost tenet."[63] But there were practical as well as ideological reasons for opposing licensing. Many, perhaps most, of the leaders of the founder societies were engineers engaged in administration or management of some sort. Since they had, to a large extent, ceased to practice engineering in a technical sense, they would not, in many cases, be eligible for licenses. If the profession were to be reconstructed around the idea of licensing, many such men would lose their influence.

The Engineers Council for Professional Development, founded in 1932, provided a focus for the debate over licensing and its alternatives. An outgrowth of a "Conference on Certification of the Profession," the ECPD sought some means of drawing a sharp line between professional engineers and other technical workers.[64] From the start the ECPD split on the means for achieving this goal. One group, led by David B. Steinman, favored universal licensing of all engineers and the formation of a closed profession in the image of the American Medical Association.[65] The leading ideological opponent of closing the engineering profession was William E. Wickenden. He thought licensing was unrealistic for engineers; cutting them off from other technical groups would ultimately produce sterility. A firm believer in individualism, Wickenden proposed an alternative plan of certification, which would mark off the successive stages in a young man's progress toward full professional standing.[66] If successful, such a scheme would mark off professional engineers from other groups without legal restrictions on admission.

In the struggle over licensing within the ECPD, the conservatives were in a dominant position. In form a joint conference committee, the ECPD included the founder societies, the Society for the Promotion of Engineering Education, the American Institute of Chemical Engineers, and the National Council of State Boards of Engineering Examiners. Practically, this organization gave a veto to each of the constituent societies. Thus, while there was substantial support for licensing, the ECPD refused to sanction any such measures.[67]

The first chairman of the ECPD, and its dominating figure in its early years, was Charles F. Hirshfeld. He made the conservative spirit of the council very clear. He held:

> We are living in the period in which it has become the style to question . . . practically everything having to do with the life of man. Religious forms and teachings, social organizations and values, economic theories and practices . . . are being subjected to an inquisition. . . . If we are, as I suspect, to return in the end to essentially what we had before, the sooner we can get a larger part of the educated and potentially powerful and influential fraction of our population thinking sane thoughts . . . the better. It is practically the function of the Council to produce just such a result with respect to some part of that fraction.[68]

The ECPD, while willing to make some minor concession, adhered firmly to Hirshfeld's conservative program. In 1931, Hirshfeld hinted at the desirability of restricting the number of students entering engineering colleges, even though he maintained that only "kindly and gentle" means would be used.[69] However, engineering educators opposed any limitations on enrollments. In 1935, the ECPD adopted a program for a system of certification parallel to licensing. But this failed to win support from the major engineering societies. Each society defined "professional" differently in its membership grades, and it proved impossible for them to reach agreement on this touchy subject.[70]

Unable to get action by the ECPD, the dissidents established their own organization in 1934, the National Society of Professional Engineers. The moving spirit in the NSPE was David B. Steinman. He favored strong licensing laws applying to all engineers.[71] Steinman also sought the "selective limitation of student enrollment," and he denounced the engineering educators for their opposition to any such proposals.[72] But Steinman was firmly opposed to collective bargaining, since it was unprofessional and discouraged "individual effort, loyalty, and ambition."[73] In form, the national society was a federation of state societies, each of which was concerned with the development of state licensing laws. Membership in the NSPE was restricted to licensed engineers. Symptomatic of the declining vitality of social responsibility among engineers was the slight emphasis the NSPE gave to public service. It was listed fourteenth in a statement of the society's objectives.[74]

The NSPE was only partially successful. Under its aegis, many states adopted licensing laws, including seven in 1935 alone; by 1947, all had enacted such legislation.[75] But licensing failed to restrict the number of engineers or close the profession. Licensing laws for engineers apply only to engineers in responsible charge of work. Although of benefit to consulting engineers, such measures do not apply to the mass of employed engineers. In any case, licensing does not get at the problems of bureaucracy and professional autonomy. By 1954, the NSPE had thirty thousand members, fewer in proportion to the total profession than the AAE had had in 1920. But if the membership of the NSPE has remained at a rather low level—generally less than 5 percent of all engineers—it has been influential nevertheless. Unlike the AAE, it has shown itself to be durable. Although small, the NSPE represented a power center outside the founder societies and an important source of agitation for professional ethics. It has functioned to some extent as the conscience of the profession.[76]

If the traditional learned professions provided one model for defending the interests of engineers, labor unions offered a second. Unions for engineers appeared in the early 1930s. But the first major thrust for unionism came after the passage of the Wagner Act in 1935. The interest was more negative than positive; under the terms of that law professional employees might be forced into heterogeneous unions against their will. The defense was to form unions for engineers.[77] But the motivation was not entirely negative. One disgruntled engineer replied to the charge that unionism was unprofessional with the query, "Is being professional and being an obsequious ass just about one and the same thing?"[78]

The mixture of motives influencing engineers on the question of unions was illustrated by the ASCE. Because of its high membership requirements, the ASCE was less hampered by business domination; at the same time, civil engineers were hard hit by the depression. In 1932, the society adopted a comprehensive model registration law.[79] The continuing discontent within the society led the ASCE to establish a committee on unionization. In 1938, the committee recommended that the society actively promote the welfare of its members in order to make unions unnecessary. Specifically, it called on the ASCE to draw up a schedule of minimum salaries for civil engineers. In 1939, the society adopted such a schedule.[80]

Measures such as licensing and salary schedules proved inadequate; and in 1943, the ASCE accepted collective bargaining. The immediate cause of this reversal in policy was the forcible unionization in that year of about one hundred engineers and architects employed at the Sunflower Ordnance Plant.[81] President Ezra B. Whitman announced that the board had taken the "momentous step" of deciding to sponsor collective bargaining.[82] But the change proved to be less lasting than anticipated. The society's aim had been chiefly preventative; when the Taft-Hartley law removed the threat of forcible inclusion in heterogeneous unions, the ASCE lost interest in collective bargaining.

Sentiment for unionization, however, gained momentum among rank-and-file engineers, and it became a major issue in the 1950s. In 1954, the ASCE polled its membership and found that 40 percent were not opposed to collective bargaining and that 25 percent were looking forward to it as a means of improving their economic status.[83] The results of a similar poll by the ASME were not made public, but according to *Business Week* the proportion of members favoring unions was about the same as for the ASCE.[84]

One reason for the shift in emphasis in the postwar period from licensing to unionization was the changing economy. Licensing had had relevance in an age of unemployment. When the postwar shortage of engineers developed, interest tended to focus on relations with employers, rather than on limiting admission to the profession. The postwar boom in college enrollments tended to swell the profession with a large number of younger men less inhibited than their elders had been in respect to unionization. One result of the new interest in collective bargaining was the formation, in 1952, of a comprehensive national organization, the Engineers and Scientists of America.

Despite the widespread discontent of younger engineers and the flurry of interest in collective bargaining, unionism failed to take deep root. Several factors contributed to the partial failure of trade unionism among engineers, but the most important was professionalism. Engineers were reluctant to combine with technicians, even though such a strategy was essential for effective organization. They were even more reluctant to collaborate with ordinary trade unions. Professionals did not like to mix with nonprofessionals. Amalgamation with labor threatened to cut off engineers not only from their professional brethren but also from management. Professional inhibitions prevented engineers from using the strike weapon effectively. As a result of factors such as these, collective bargaining tended to lose ground among engineers in the 1960s.[85]

Although the movements for licensing and collective bargaining for engineers did not gain their maximum strength until after the Second World War, their appearance in the 1930s was significant. They indicated a growing split between professionalism and social responsibility. This placed the American Engineering Council in jeopardy. The very rationale for its existence was disappearing. The end came in 1940, when the founder societies abolished the AEC.

The end of the AEC may have been hastened by its alignment with business. After 1935, the council, like much of the business community, was frankly opposed to the New Deal. This circumstance naturally reduced the effectiveness of the council in dealing with government agencies.[86] The council had nothing to say that business was not saying with greater authority. Indicative of the ideological *cul-de-sac* into which the council had fallen by the later 1930s was the fact that the council refused to comply with the provisions of the Social Security Act, and in 1939, had to pay three years' back taxes plus penalties and interest.[87] This gesture may have pleased right-wing businessmen, but it had little relevance to the profession of engineering.

Perhaps even more serious for the council's future was its inability to advance the status and the welfare of engineers. While the council spent a good deal of time on such trivial matters as the telephone book classification of engineers, it was unable to take any significant action to grapple with the problems that beset the profession.[88] In 1929, the council authorized a "Committee on Engineering and Allied Technical Professions," whose aim was "to enable the Council to enter upon a program designed to improve the general status of the engineering profession."[89] The committee conducted a large statistical study of the engineering profession. The council took no significant action, however. One reason for this was the hostility of the council's president, Arthur W. Berresford, to any collective action. He approved the study because he thought it would vindicate the proposition that each individual gets what he deserves. In order to square this belief with the findings of the committee, he had to argue that pay consists of nonmonetary rewards as well as of hard cash.[90] The council favored improving the position of engineers in government employment, but only if it could do so "without becoming involved in anything leading toward the socialization of engineering or the standardization of either duties or salaries which might limit engineering initiative."[91]

By the late 1930s, the AEC no longer had substantial support from the rank-and-file of the engineering profession. The final blow came from the ASCE, which had adopted, by 1938, an activist program of professional welfare that was completely out of harmony with the policies of the AEC. The ASCE suggested the establishment of a joint conference committee to determine the fate of the council.[92] In 1939, Vannevar Bush made an eloquent plea for the continuation of the AEC and its ideal of "ministering to the people."[93] But he was almost alone; most engineers no longer thought of social responsibility and professionalism as fundamentally linked. In any case, the council had demonstrated its incapacity to advance either end. The ASCE and the AIEE decided to withdraw further support, and the council was abolished at the end of 1940.[94]

NOTES

1. For an indication of the public interest in national planning in the early days of the depression, see Charles A. Beard and Mary R. Beard, *America in Midpassage*, 2 vols. (New York, 1939), I, 98–111, *passim*.

2. For Veblen, see Edwin T. Layton, "Veblen and the Engineers," *American Quarterly*, XIV (Spring, 1962), 64–72. For Beard, see his *America in Midpassage*, II, 839–848, and "Government by Technologists," *New Republic*, LXIII (June 18, 1930), 115–120. For a suggestive interpretation, see also Cushing Strout, "The Twentieth-Century Enlightenment," *The American Political Science Review*, XLIX (June, 1955), 321–339.

3. For Wallace, see his "The Engineering-Scientific Approach to Civilization," *Mechanical Engineering*, LVI (March, 1934), 131–134.

4. C. A. Norman, "Industrial Fundamentals," *The Journal of Engineering Education*, XXII (March, 1932), 537–542.

5. Edwin T. Layton, "Veblen and the Engineers," *American Quarterly*, XIV (Spring, 1962), 64–72.

6. *Ibid.* See also Joseph Dorfman, *Thorstein Veblen and His America* (New York, 1934), 459–460, 462, 510–514.

7. Henry Elsner, Jr., *The Technocrats, Prophets of Automation* (Syracuse, New York, 1967), 2–3.

8. *Ibid., passim.*

9. Walter Rautenstrauch, "Public Enterprise" *Electrical Engineering*, LII (April, 1933), 234–236, and Bassett Jones, "Engineers Versus Business Men and Politicians" (letter to the editors), *Electrical Engineering*, LII (June, 1933), 430–431.

10. Walter Rautenstrauch, "The Problems of Public Enterprise," *Mechanical Engineering*, LV (March, 1933), 150. Rautenstrauch later reverted to a conventional progressivism (see his *Who Gets the Money?* [New York, 1939], *passim*).

11. "American Engineering Council," *Electrical Engineering*, LII (February, 1933), 137–138.

12. For the engineers' reaction to technocracy, see G.A.S., "Bardell versus Pickwick—A Review of Technocratic Literature," *Mechanical Engineering*, LV (March, 1933), 203–206.

13. H. P. Charlesworth, "The Engineer and a Century of Progress," *Electrical Engineering*, LII (July, 1933), 445.

14. Bancroft Gherardi, "Engineer and Progress," *Electrical Engineering*, LII (April, 1933), 257–258.

15. W. S. Lee, Roy V. Wright, Robert E. Tally, Francis Lee Stuart, and H. Hobart Porter, "Engineering and Human Happiness," *Electrical Engineering*, L (August, 1931), 641–642.

16. C. E. Kenneth Mees, "Has Man Benefitted by Engineering Progress," *ibid.*, 642–649.

17. C. F. Hirschfeld, "Whose Fault?" *Mechanical Engineering*, LIV (March, 1932), 180.

18. Roy V. Wright, "The Engineer Militant," *Mechanical Engineering*, LIV (January, 1932), 20–22.

19. Frederick M. Becket, "New Responsibilities of the Engineer," *Mining and Metallurgy*, XV (March, 1934), 130.

20. William S. Lee, "Coordination—The Essence of Modern Engineering," *Electrical Engineering*, L (July, 1931), 513.

21. William S. Lee, "The Engineer's Duty to Himself and to the Public," *Electrical Engineering*, L (February, 1931), 125.

244   THE REVOLT OF THE ENGINEERS

22. Arthur W. Berresford, "Progress is the Way Out of the Depression," *Electrical Engineering*, L (December, 1931), 947.

23. "Engineering Council Adopts Modification of Its Constitution," *Engineering News-Record*, CII (January 17, 1929), 117, and American Engineering Council, Annual Meeting Proceedings, January 10–11, Appendix E, pp. 1–2, American Engineering Council Papers, Engineering Societies Library, New York (hereafter cited as AEC Papers).

24. American Engineering Council, Efforts to Increase the Economic Status and Employment of Engineers (mimeograph), June 6, 1932, pp. 9–10, AEC Papers.

25. "The Balancing of Economic Forces," *Mechanical Engineering*, LIV (June, 1932), 423.

26. American Engineering Council Miscellaneous Reports, Executive Committee Meeting Minutes, April 27, 1936, p. 4, AEC Papers.

27. "American Engineering Council and Public Construction," *Engineering News-Record*, CIX (December 8, 1932), 689–691, and "As to Credit Where Due," *ibid.*, 692.

28. Roy V. Wright, "The Engineer's Status in the Community," *Mechanical Engineering*, LVII (June, 1945), 401–402, 406.

29. "Report of Institute Committee on Sponsoring Discussions of Social and Economic Subjects," *Electrical Engineering*, LIV (June, 1935), 672–673.

30. William H. Wood, "A Forum for Social Problems" (letter to the editors), *Mining and Metallurgy*, XVII (February, 1936), 105.

31. Marion B. Richardson, "Aims and Objects of the A.S.M.E." (letter to the editors), *Mechanical Engineering*, LXIV (March, 1942), 232.

32. Ralph E. Flanders, "The Work and Influence of the Engineer," *Mechanical Engineering*, LII (September, 1930), 829.

33. Ralph E. Flanders, "The Engineer's View," *Mechanical Engineering*, LIII (February, 1931), 99, 103.

34. *Ibid.*, 103.

35. Ralph E. Flanders, "An End to Unemployment," *Mechanical Engineering*, LVI (September, 1934), 519.

36. Ralph E. Flanders, "Neglected Elements of Recovery," *Mechanical Engineering*, LVII (June, 1935), 347.

37. *Ibid.*, 348.

38. Ralph E. Flanders, *Platform for America* (New York, 1936), *passim*.

39. Ralph E. Flanders, "New Pioneers on a New Frontier," *Mechanical Engineering*, LVIII (January, 1936), 4.

40. Ralph E. Flanders, "The Engineer in a Changing World," *Electrical Engineering*, LVI (August, 1937), 937–941.

41. Ralph E. Flanders, "Progress Report of an Amateur Economist," *Mechanical Engineering*, LXII (July, 1940), 527.

42. William E. Wickenden, "The Engineer in a Changing Society," *Electrical Engineering*, LI (July, 1932), 467.

43. William E. Wickenden, "The Engineer and the New Deal," *Electrical Engineering*, LII (September, 1933), 598.

44. William E. Wickenden, "The Engineer in a Changing Society," *Electrical Engineering*, LI (July, 1932), 467.

45. William E. Wickenden, "The Social Sciences and Engineering Education," *Mechanical Engineering*, LX (February, 1938), 150.

46. William E. Wickenden, "The Young Engineer Facing Tomorrow," *Mechanical Engineering*, LXI (May, 1939), 347.

47. *Ibid.*

48. *Ibid.*, 348.

49. William E. Wickenden, "Engineering Education Needs a 'Second Mile,' " *Electrical Engineering*, LIV (May, 1935), 471–473.

50. Arthur E. Morgan, "Social and Economic Implications of TVA," *Civil Engineering*, V (December, 1935), 754, 757.

51. David E. Lilienthal, *The Journals of David E. Lilienthal*, 2 vols. (New York, 1962), I, *The TVA Years*, 42–43.

52. Arthur E. Morgan, *Design in Public Business* (Yellow Springs, Ohio, 1939), 30. See also Lilienthal, *The TVA Years*, 39, 61–63, 70–76.

53. For Morgan's views on autonomy, see Arthur E. Morgan, "Engineer's Share in Democracy," *Civil Engineering*, IX (November, 1939), 637–638.

54. Arthur E. Morgan, *The Small Community, Foundation of Democratic Life* (New York, 1942), 18.

55. *Ibid.*, 74–75.

56. *Ibid*, 78–79.

57. Arthur E. Morgan, *Edward Bellamy* (New York, 1944), *passim*. See also Arthur E. Morgan, *Nowhere Was Somewhere* (Chapel Hill, 1946), *passim*.

58. James W. Parker, "The Spirit of a People," *Mechanical Engineering*, LXV (January, 1943), 6.

59. Malcolm Pirme, "Chart for the Future," *Trans ASCE*, CIX (1944), 1430–36.

60. For a survey of the postwar reaction, see Eric F. Goldman, *The Crucial Decade* (New York, 1956), 8, *passim*.

61. Robert K. Burns, "The Comparative Economic Position of Manual and White-Collar Employees," *Journal of Business*, XXVII (October, 1954), 258–261.

62. Frederick H. McDonald, "The New Competition and New Horizons in Engineering," *Civil Engineering*, I (December, 1931), 1361–63.

63. D. M. Liddell, "Licensing Engineers Not Analogous to Licensing Doctors" (letter to the editors), *Mining and Metallurgy*, III (February, 1922), 35.

64. "Engineers Council for Professional Development," *Civil Engineering*, II (August, 1932), 515, and R. I. Rees, "The Professional Development of the Engineer," *Electrical Engineering*, LII (February, 1933), 129–130.

65. "Engineers Council for Professional Development Discussed," *Electrical Engineering*, LII (December, 1933), 932–933, and D. B. Steinman, "Registration of Engineers" (letter to the editors), *Electrical Engineering*, LV (July, 1936), 844–845.

66. William E. Wickenden, "Engineering Education Needs a 'Second Mile,' " *Electrical Engineering*, LIV (May, 1935), 471–473.

67. "Engineers Council for Professional Development," *Civil Engineering*, II (August, 1932), 515.

68. "Engineers' Council for Professional Development Discussed," *Electrical Engineering*, LII (December, 1933), 932.

69. C. F. Hirshfeld, "Engineers Council for Professional Development," *Mechanical Engineering*, LIV (September, 1932) 643.

70. "The E.C.P.D. Program to Gain Recognition for the Engineering Profession," *Electrical Engineering*, LIV (July, 1935), 786–787; J. F. Fairman, "Relation of ECPD to Professional Societies Discussed," *Electrical Engineering*, LX (December, 1941), 606–607.

71. Paul H. Robbins, "National Society of Professional Engineers," *General Electric Review* (March, 1954), 47–49, and D. B. Steinman, "Registration of Engineers" (letter to the editors), *Electrical Engineering*, LV (July, 1936), 844–845.

72. D. B. Steinman, "Engineering Education," *American Engineer*, VI (July–August, 1936), 9.

73. D. B. Steinman, "Letters and Comments" (letter to the editors), *Mechanical Engineering*, LXII (May, 1940), 246.

74. D. B. Steinman, "What the National Society of Professional Engineers Can Accomplish," *American Engineer*, V (January, 1935), 6, 14, 19.

75. E. W. Ellis, "Licensing and Registration of Engineers in the United States," *Mining and Metallurgy*, XXVI (January, 1945), 22.

76. Paul H. Robbins, "National Society of Professional Engineers," *General Electric Review* (March, 1954), 47–49.

77. National Society of Professional Engineers, *A Professional Look at the Engineer in Industry*, (Washington, D. C., 1955), 56–59, and Herbert N. Northrup, *Unionization of Professional Engineers and Chemists* (New York, 1946), 10–12.

78. David B. Ericson, "Closing the Discussion on the Underpaid Young Engineer" (letter to the editors), *Mining and Metallurgy*, XXIII (September, 1942), 474.

79. "A Model Law for the Registration of Professional Engineers and Land Surveyors," *Civil Engineering*, II (August, 1932), 517–520.

80. "Special Committee on Unionization Reports," *Civil Engineering*, VIII (March, 1938), 216–217; "Civil Engineers' Salaries," *Civil Engineering*, IX (September, 1939), 566–568.

81. "Engineers Protest Affiliation with Sub-Professionals," *Civil Engineering*, XIII (July, 1943), 337–338, and "Board Actions on Collective Bargaining," *Civil Engineering*, XIII (September, 1943), 443.

82. Ezra B. Whitman, "Collective Bargaining and Salaries for Professional Engineering Employees," *Civil Engineering*, XIII (November, 1943), 513–514, and "Professional Engineers in Southern California form Bargaining Units," *Civil Engineering*, XVI (May, 1946), 213–215.

83. "Collective Opinion of Members on Collective Bargaining," *Civil Engineering*, XXIV, pt. 1 (May, 1954), 317–319, and Charles W. Yoder, "What About Collective Bargaining," *Civil Engineering*, XXIV, pt. 2 (September, 1954), 591–593.

84. "Engineering Associations Take on a Union Patina," *Business Week* (August 28, 1954), 110.

85. Richard E. Walton, *The Impact of the Professional Engineering Union* (Boston, 1961), 18–45.

86. Frederick M. Feiker to Professor John S. Dodds, November 10, 1938, AEC disbandment volume, AEC Papers.

87. American Engineering Council, Finance, 1919–1956, financial statement for year 1939, AEC Papers.

88. American Engineering Council, Reports of Executive Secretary, vol. IV, January 13–14, 1933, pp. 9–10, and January 12–13, 1934, pp. 15–20, AEC Papers.

89. "Council Appoints Committee on Engineering and Allied Technical Professions," *Journal AIEE*, XLVIII (July, 1929), 567.

90. "Arthur W. Berresford Gives Message to Engineers," *American Engineering Council Bulletin*, IX (February, 1930), 2, 6.

91. "The News From Washington," *Mechanical Engineering*, LIX (July, 1937), 564.

92. "Meetings of Outgoing Board of Direction—Secretary's Abstract," *Civil Engineering*, IX (March, 1939), 189.

93. Vannevar Bush, "The Professional Spirit in Engineering," *Mechanical Engineering*, LXI (March, 1939), 198.

94. "Current Items from American Engineering Council," *Electrical Engineering*, LIX (January, 1940), 45, and "Societies Terminate Affiliation with A.E.C.," *Electrical Engineering*, LX (January, 1941), 31. See also American Engineering Council, Joint Committee on Correspondence, 1938–1940, *passim*, AEC Papers.

EPILOGUE

THE RISE OF SCIENTIFIC PROFESSIONALISM

The abolition of the American Engineering Council produced scarcely a ripple. By 1940, few engineers were even aware of the existence of the AEC; even fewer lamented its demise. In any case its functions did not lapse completely. In the following year, the founder societies organized a joint conference committee to consider common problems. During the war it served the government as the Engineers Defense Board, although the scope of its activities appears to have been more restricted than that of the Engineering Council in the First World War. In 1945, the name was changed to the Engineers Joint Council. In 1949, the EJC adopted a constitution and sought to recruit new members. Many engineering societies affiliated themselves with the EJC. There were, however, two conspicuous exceptions. The NSPE objected, in principle, to any unity organization not based on individual memberships. The IRE continued to take the position that it was not competent to represent its members in professional and political matters. But behind both rejections lay a fundamental difference. The NSPE and the IRE stood for an independent profession; the EJC was committed to an alliance with business.[1]

Although nominally devoted to the defense of a profession, the EJC showed a great concern for the interests of employers of engineers. It helped to secure the modification of the Wagner Act so as to prevent the forcible unionization of professional employees. It played an important role in stemming the tide of unionism among engineers in the 1950s. In the case of licensing, the EJC, like it predecessors, took an indirect approach. It did not oppose licensing in principle, but it attempted, successfully, to prevent such measures from significantly restricting admission to the engineering profession. Indeed, one of the EJC's principal concerns was that of increasing the supply of engineers. The alleged shortage of engineers in the 1950s, however, was a matter of debate. There unquestionably was a boom in advertising for engineers by defense industries where the government paid the advertising bills. While engineers' salaries increased, there was evidence that the rise was less than that of most other professions.[2]

In its struggle to keep the engineering profession open, the EJC was continuing policies begun by the Engineering Council and the AEC. But in at least one area the EJC initiated significant new departures. It took important steps to promote close cooperation between engineering and the scientific community. The EJC was responsible, in part, for the elevation of engineering to divisional status within the National Science Foundation, in 1964, and the appointment in the same year of a past-president of the EJC to the Science Board of the NSF. A more momentous step was the establishment of a National Academy of Engineering, in 1964, under the charter of the National Academy of Sciences. The initiative came from the president and other principal leaders of the EJC. This move enabled engineers to capitalize on the advisory position that scientists had established over the years. Government agencies were quick to seek the advice of the new organization. In general, the academy has found that requested projects have grown more rapidly than the organization's ability to deal with them.[3]

The National Academy of Engineering has not been in existence long enough for historical evaluation. But there is some question as to whether it can evolve into an autonomous professional agency. The original membership was drawn from the EJC. New members are elected by existing members. This creates a danger that the academy membership will reflect the same business orientation as the EJC. In fact, ties of company loyalty have not been insignificant factors in the selection of new members, and presidents and other executive officers of large corporations are conspicuous on its roster. Finances present another delicate problem. The dues have been modest and quite insufficient to sustain the ambitious program of the academy. The academy has relied on large gifts from the corporation executives among its members, as well as the financial support of private business. Whether this dependence on business for funding will influence the policy recommendations of the academy is an open question. The prior experience of American engineering societies, however, is far from reassuring on this score.[4]

The shift toward scientific professionalism was not simply a calculated strategy by American engineers. In a deeper sense, engineers were being drawn into the orbit of the scientific community by the forces of continuing technological change. Engineering as a mass profession had been brought into existence by a scientific revolution in technology. But down to 1945, this revolution was still in its infancy. Since 1945, the scientific revolution in technology has progressed even further and produced significant institutional changes. These are perhaps best seen in engineering education. The scientific content of the curriculum has been greatly increased. This has consisted not merely of students taking more science, but of increasing the scientific content of the engineering courses. One symptom of this has been the recruitment by engineering faculties of holders of Ph.D. degrees in physics and other sciences. Conversely, the element of art in engineering has declined; "practical" training in shop and field has virtually ceased. The four-year curriculum has become increasingly inadequate, and the number of engineers receiving advanced degrees has steadily grown. There has been much talk of making the master's the first professional degree, for there is much evidence, in the twenty years or so after 1945, of a quantum jump in engineering knowledge.[5]

In a few fields the transformation of engineering took place much earlier. Radio engineering had been intimately associated with science from the outset. This had a profound influence on the IRE. Alone among American engineering societies, the IRE adopted a strategy of scientific professionalism. This was a major factor in causing the IRE to stand aloof from the professional activities of the founder societies. The IRE was, therefore, in an excellent position to capitalize on the continuing scientific revolution in technology. It absorbed the new field of electronics, and its growth was spectacular. By the 1960s, the IRE had outgrown the AIEE in both quantity and quality. In 1963, the two societies merged to form a new organization, the Institute of Electrical and Electronic Engineers, or IEEE. At the time of the merger, the IRE had a membership in excess of 100,000, more than twice that of the AIEE.[6]

The IEEE was not only the largest American engineering society; it received from the IRE a strong tradition of professional independence. From the AIEE it inherited a membership in the EJC. This uneasy combination came to a crisis in 1967 when the EJC adopted a new constitution, a salient feature of which was to allow corporations to become members. While this innovation provided a new source of funds, overt business support seriously compromised the EJC's claim to speak for an autonomous profession. The response of the IEEE was prompt; it announced that it was withdrawing from the EJC at the end of the year. The loss of the IEEE was a serious blow to the EJC and to the strategy of alliance with business for which it stood.[7]

As engineers were drawn increasingly into intimate collaboration with the scientific community, they tended, ironically, to reject the ideology of engineering. However, this loss of interest in science as a source of identification and role definition was not without precedent. The IRE—within the most scientific of engineering fields, radio engineering —had never shown an interest in this ideology or in the professional activities based on it. In other engineering fields a loss of enthusiasm became apparent in the 1930s. By the 1960s, disenchantment was being replaced in some cases by outright repudiation. A few engineers saw science as a threat to the identity of their profession. One engineer attacked the "science worship syndrome" and maintained that engineering was broader and more challenging than science.[8]

It was perhaps paradoxical that as engineers became more scientific they tended to reject science, at least as a means of defining themselves and their social role. The explanation lies in certain of the practical consequences of the continuing scientific revolution in technology. As engineers came to work in the same units with scientists, two facts became apparent. In the scientific "peck order," engineers stood at the bottom. The highest prestige went to the more abstract and theoretical fields. Secondly, beginning with the early industrial research laboratories, scientists have tended to displace engineers in the vanguard of technological change. In consequence, scientists have often received the glamour and the credit, especially in such newer and more rapidly growing areas as nuclear energy, electronics, and aerospace. It would not be surprising if engineers rejected an identification that denied them prestige and a stellar role in technological progress.

The decline in the ideology of engineering may have opened the way for the development of a new interest in the social effects of technology. The ideology of engineering emphasized social responsibility, but it also served to prevent effective action. Like a set of blinders for the profession, this ideology diverted the genuine concern of some engineers with the misuse of technology into unproductive channels. The ideology of engineering, with its stress of the superior, if not superhuman, qualities of the engineer, encouraged sterile status seeking and prestige politics. It contained a large element of fantasy and wish-fulfillment, which hindered realistic action. In particular, this ideology committed the profession to finding engineering solutions to social problems. Thus, it encouraged engineers to substitute an impossible task for one that was merely difficult. A further liability was that it led many engineers into the blind alley of technocracy. Whether engineers will gain a renewed social awareness after being freed of this incumbrance is a moot question. If they do, a new engineering progressivism is not beyond the realm of possibility.

NOTES

1. "Engineers Joint Council," *Mechanical Engineering*, LXXII (May, 1950), 399, 402, and T. A. Marshall, Jr., "Engineers Joint Council," *Mechanical Engineering*, LXXVI (July, 1954), 582–584.

2. On the shortage of engineers, see David M. Black and George J. Stigler, *The Demand and Supply of Scientific Personnel* (New York, 1957), 24–32.

3. Eric A. Walker, "The Work of the National Academy of Engineering, U.S.A.," *The Journal of the Royal Society for the Encouragement of Arts, Manufactures and Commerce*, CXVII (May, 1969), 385–390.

4. *Ibid.*, 389, 391, 393.

5. Gordon S. Brown, "New Horizons in Engineering Education," *Daedalus*, XCI (Spring, 1962), 341–361, and R. B. Adler, "Science and Engineering Education," *Journal of Engineering Education*, XLVII (October, 1956), 121–128.

6. Special Merger Supplement," *Electrical Engineering*, LXXXI (April, 1962), and B. R. Teare, Jr., "Message from the President," *Electrical Engineering*, LXXXI (December, 1962), 915.

7. "IEEE Withdraws from Membership in EJC," *IEEE Spectrum*, IV (November, 1967), 20, and Stan Klein, "Togetherness Flops at EJC," *Machine Design*, XL (February 1, 1968), 8, 10. The American Institute of Chemical Engineers also withdrew from the EJC.

8. W. R. Marshall, Jr., "Science Ain't Everything," *Chemical Engineering Progress*, LX (January, 1964), 17–21.

BIBLIOGRAPHIC ESSAY

PRIMARY SOURCES

The most important primary sources used in this study were the manuscripts and the engineering journals. In sheer volume and quantity of information, the journal literature was by far the greater. Manuscripts, however, served an indispensable function in revealing behind-the-scenes maneuvers and in providing insights into ideas and motives.

Manuscripts

Four manuscript collections were particularly important because the men involved were principal actors in the story I have to tell. They were: the Frederick William Taylor Collection at the Stevens Institute of Technology, Hoboken, New Jersey; the Papers of Morris L. Cooke at the Franklin D. Roosevelt Presidential Library, Hyde Park, New York; the Herbert Hoover Papers at the Herbert Hoover Presidential Library, West Branch, Iowa; and the Papers of Frederick H. Newell at the Manuscripts Division of the Library of Congress.

The Taylor Papers were the most helpful. They were very complete, not only for Taylor's own activities, but for those of his associates as well. For the period from 1903 to 1915, they are a better source for Cooke's activities than the Cooke Papers. The Cooke Papers, although voluminous, are less complete. Most of the material is dated after 1915. The Cooke collection includes a complete set of his published papers, including the anonymous ones. I was fortunately able to supplement these sources by personal interviews with Cooke on June 27 and 28, 1957, and by an extensive personal correspondence with him, which extended over several years.

The Papers of Frederick Haynes Newell, even though fragmentary and containing little correspondence, include a pocket diary he kept throughout his career. The notes are cryptic and exclude motives, thoughts, and plans. But they do cover his movements, the people he talked with, and his major activities. Also of great value were the scrapbooks of newspaper and magazine clippings covering his association with the Reclamation Service. They provide insights into the political battles that led to the eventual dismissal of Newell from the government. The papers include typescripts of many of Newell's addresses; in several cases these include materials deleted from the printed versions that appeared in the engineering press. Also included is an imperfect stenographic transcript of the third conference on cooperation.

The Herbert Hoover Papers are particularly rich in material relating to the formation of the FAES and its subsequent transformation into the AEC. This collection also includes much information on the internal politics in the AIME from 1919 to 1925. Among the most useful sections of these papers were the manuscript records of the Committee on the Elimination of Waste of the FAES. The Hoover collection includes a complete set of Hoover's speeches and public statements. These are useful because Hoover's speeches were reprinted under a variety of titles and in various degrees of completeness.

Several other manuscript collections were of value for particular episodes. The Papers of Louis D. Brandeis at the School of Law Library of the University of Louisville contain information on Brandeis's involvement with scientific management and with Morris L. Cooke in the battle against the utilities. Included is some useful material on Cooke's ethics "trial" before the ASME. The Papers of Frank B. Gilbreth in the Industrial Relations Library of Purdue University provide information on the scientific-management movement and some of Cooke's reform ventures. The Ralph D. Mershon Collection at Ohio State University contains little correspondence, but it provides some information on Mershon's clashes with the utilities. The William Smyth Collection at the Bancroft Library of the University of California, Berkeley, is chiefly the accumulation of a lonely eccentric who coined the word "technocracy." The William E. Wickenden Papers in the Archive of Contemporary Science and Technology of Case Western Reserve University provide only occasional glimpses of Wickenden's leadership of the profession and the AIEE in the 1930s.

Thorstein Veblen's activities at the New School in New York after the First World War led indirectly to the technocracy movement of the 1930s. The papers of Guido H. Marx in the possession of Mrs. Barbara Givan, Palo Alto, California, include correspondence with Cooke and Veblen relative to the course that Veblen persuaded Marx to give at the New School. They may be supplemented by the Marx file in the Cooke

Papers. The Otto S. Beyer Papers at the Manuscripts Division of the Library of Congress include some material on Beyer's efforts to continue Marx's work as well as his activities on behalf of labor unions for engineers in the 1920s and 1930s.

Two engineers allowed me access to their private collections. Bernhard F. Jakobsen sent me materials from his personal files relating to his ethics case before the ASCE. In the course of our correspondence he provided much information on the ethical standards in civil engineering over the years, including some interesting information about Frederick Haynes Newell. Professor Julian W. King of the University of California allowed me to see some of his correspondence relating to censorship in the ASME in the 1950s.

The American Engineering Council Papers at the Engineering Societies Library, New York, contain much of value on the Engineering Council, the Federated American Engineering Societies, and the American Engineering Council. Most of the material consists of official minutes, reports, and the like, but some correspondence is included on the dissolution of the AEC. Of particular value is a stenographic transcript of the conference that created the FAES.

Engineering Periodicals

Although the engineering periodicals are exceedingly valuable, their use involves many pitfalls and difficulties. First, there is the sheer difficulty of finding materials. Some of the most useful information on social and professional topics came from letters to the editors, editorials, brief news items, and the like. This information is not indexed in guides like *Engineering Index* and *Industrial Arts Index*. Secondly, *Engineering Index* did not begin its author index until 1928; prior to that time it is difficult to run down all of an author's articles. The investigator must therefore resort to the examination of runs of individual periodicals. Most engineering journals have annual indexes and some have cumulative indexes. But social and professional materials may be left out or hard to find. Therefore, I found the best method to be the tedious one of going through the journals page by page. Because of the bulk of the materials and the time involved, this could be done for only a few periodicals at particularly critical periods.

The most important journals were the periodicals published by the major national societies: the American Society of Civil Engineers, the American Society of Mechanical Engineers, the American Institute of Electrical Engineers, and the American Institute of Mining Engineers.

Each publishes at least two different serials, typically a *Proceedings* and a *Transactions*. The *Proceedings* of the ASCE, ASME, and AIEE and the *Bulletin* of the AIME evolved into monthly news magazines: *Civil Engineering, Mechanical Engineering, Electrical Engineering,* and *Mining and Metallurgy.* The magazines are generally much richer in society news and information on the profession than are the *Transactions.* The latter, however, include presidential addresses, discussions of papers, obituary biographies, and committee reports. In addition, each society publishes separate yearbooks and other materials.

The publications of the smaller national and local engineering societies usually are useful only for their own history; they contain little relating to the profession as a whole. There are three important exceptions, however. *The Journal of Engineering Education* served as an important forum of opinion on social and professional matters. Both of the major national protest organizations published journals. *The Monad* —later *Professional Engineer*—was the organ of the American Association of Engineers. *American Engineer* served the same function for the National Society of Professional Engineers.

Among the publications of other engineering societies, the Institute of Radio Engineers' *Proceedings* are in a special category for two reasons. The IRE eventually outgrew and absorbed the AIEE. The IRE also considered itself, not without justice, to represent a higher level of professional development than the founder societies. Among other engineering society periodicals consulted were the *Bulletin of the Taylor Society,* the *Proceedings of the Mining and Metallurgical Society of America,* the *Journal of the Society of Automotive Engineers,* the *Transactions of the Illuminating Engineering Society,* the *Transactions of the American Society of Heating and Ventilating Engineers,* the *Transactions of the American Electrochemical Society,* and the *Journal of the Cleveland Engineering Society.*

The publications of engineering societies suffer from one severe drawback: professional associations attempt to keep their own inner workings secret. The commercial journals of engineering fill in this gap, to some extent, and include much other information on professional development. Of particular value were the journals whose editors were major figures in efforts to reform the profession. A classic example, which is of particular value because it attempted to cover all fields of engineering, was *Engineering News*—later *Engineering News-Record.* Under the editorship of Charles Whiting Baker, it was a major force of reform in the entire profession. In the 1920s it gradually lost its reforming zeal and evolved into a journal of civil engineering; as such, it con-

tinued to publish material on the ASCE. It may be supplemented by
Engineering Record for the period prior to the merger of these two
periodicals in 1917.

Not all editors were reformers. The major publications in mining and
electrical engineering, *The Engineering and Mining Journal* and *Electrical World*, tended to be allied with the dominant groups within the corresponding societies, the AIME and AIEE. Both journals contain much
information, but they must be supplemented by others with less inhibited publication policies. Smaller journals in the middle west or far west
were likely to be sympathetic to dissenting voices within their professional fields. An outstanding example is provided by *Mining and Scientific Press*, which Thomas A. Rickard edited in San Francisco and which he
made into an important vehicle for dissenting elements within the
AIME. Similarly, *The Electrical Review and Western Electrician* often
printed material ignored by *Electrical World*. For mechanical engineering, the publications of the ASME may be supplemented by *Power* and
American Machinist. The leader of the dissidents in the ASME, Morris
L. Cooke, did not utilize the commercial press extensively. Having been
associated with the printing industry, he found it easy to print his own
materials, including a journal, *Utilities Magazine*, which was vigorous,
if short-lived.

SECONDARY SOURCES

The Revolt of the Engineers, especially the first two chapters, relied
heavily on statistical studies of the number and other characteristics of
engineers. Among the best sources were publications of the United
States government. The separate censuses are not comparable with one
another. Reasonable approximations of the number of engineers over
time, along with a discussion of the problems involved in determining
the numbers, may be found in U.S. Bureau of the Census, *Sixteenth
Census, Population, Comparative Occupational Statistics for the United
States, 1870–1940* (Washington, D.C., 1943). More recent estimates:
U.S. Congress, Joint Committee on Atomic Energy, *Engineering and
Scientific Manpower in the United States, Western Europe, and Soviet
Russia*, 84th Congress, Second Session (Washington, D.C., 1956); National Science Foundation, *Scientists, Engineers, and Technicians in the
1960s, Requirements and Supply*, NSF 63–34 (Washington, D.C., 1963);
and National Science Foundation, *Scientific and Technical Manpower
Resources*, NSF 64–28 (Washington, D.C., 1964).

The Department of Labor has done several studies of the employment and earnings of engineers. These include additional data of value on education and other characteristics of engineers. They include: Andrew Fraser, *Employment and Earnings in the Engineering Profession, 1929–1934*, U.S. Department of Labor Bulletin, no. 682 (Washington, 1941), and U.S. Department of Labor Bureau of Labor Statistics, *Employment Outlook for Engineers*, Department of Labor Bulletin, no. 968 (Washington, 1949). A parallel study by the profession itself is Andrew Fraser, *The Engineering Profession in Transition* (New York, 1947).

For understanding the social characteristics of engineers, however, one work stands in a class by itself: Society for the Promotion of Engineering Education, *Report of the Investigation of Engineering Education, 1923–1929*, 2 vols. (Pittsburgh, 1934). It contains a wealth of material, including data on class and ethnic origins, education, and other topics. Although more limited in scope, a recent study is of great value: Robert Perrucci, William K. Le Bold, and Warren E. Howland, "The Engineer in Industry and Government," *Journal of Engineering Education*, LVI (March, 1966), 237–274.

Sociologists, economists, psychologists and others have analyzed the profession. Esther Lucile Brown, *The Professional Engineer* (New York, 1936), is a pioneering effort still of value. A basic source for the increasing role of the engineer in management is Mabel Newcomer, *The Big Business Executive* (New York, 1955). Newcomer's work has been extended by Jay M. Gould, *The Technical Elite* (New York, 1966). Harriet B. Moore and Sidney J. Levy, "Artful Contrivers: A Study of Engineers," *Personnel*, XXVIII (September, 1951), 148–153, is a highly suggestive psychological study. An economic study that I found useful was Robert K. Burns, "The Comparative Economic Position of Manual and White-Collar Employees," *The Journal of Business*, XXVII (October, 1954), 257–267. It is notable that the periods of greatest discontent within the engineering profession coincide with those in which Burns found that the salary differential between manual and white-collar workers was least.

Although quantitative studies were very helpful, I found some useful insights in more speculative essays. Herbert A. Shepard, "The Engineer and His Culture," *Explorations in Entrepreneurial History*, IV (May, 1952), 211–218, suggests that the engineer is a marginal man between science and business, an idea I found to be most productive. Another idea that I found useful was that ideology plays a vital role in the development of identification with an occupation, discussed in Howard Becker and James W. Carper, "The Development of Identification with an Occupation," *American Journal of Sociology*, LXI (January, 1956), 289–298.

I gained a number of ideas from two branches of sociology, the studies of bureaucracy and of professions. I was influenced by Everett C. Hughes, *Men and their Work* (Glencoe, Illinois, 1958), which stresses the continuity between professional and nonprofessional occupations. Formal models of professions I found to be of less value, since engineers lack consensus on such fundamentals as licensing and ethics. In the case of bureaucracy, I found the idea of orientation to be very helpful. William Kornhauser, *Scientists in Industry* (Berkeley and Los Angeles, 1962), summarizes this question and provides information on conflicts between professional and organizational loyalty. A recent discussion of the literature on bureaucratic orientation is Eugene Uyeki, "Behavior and Self-Identity of Federal Scientist-Administrators," in H. D. Lerner, ed., *Proceedings of the Conference on Research Program Effectiveness* (New York, 1966), 497–498. I did not apply this concept to individuals. But I did find it a key to understanding the multiplicity and relations of engineering societies.

Studies which explore the political behavior of engineers include: Robert K. Merton, "The Machine, the Worker, and the Engineer," *Science*, ser. 2, CV (January 24, 1947), 79–84; Max Lerner, "Big Technology and Neutral Technicians," *American Quarterly*, IV (Summer, 1952), 99–109; and Richard L. Meier, "The Origins of the Scientific Species," *Bulletin of the Atomic Scientists*, VII (June, 1951), 169–173. While suggestive, all should be used with care. Their assumption that the engineer is neutral or indifferent to social problems should be compared with L. W. Wallace and J. E. Hannum, "Engineers in American Life," *Mechanical Engineering*, LI (December, 1929), 899–904, a statistical study of engineers listed in *Who's Who in America*.

Only in recent years have historians viewed the engineering profession as a suitable subject for serious investigation. Much of the story must be assembled, often by inference, from the journal literature, supplemented by general histories of American technology and biographical studies. An indispensable beginning for investigation is Eugene S. Ferguson, *Bibliography of the History of Technology* (Cambridge, 1968). Melvin Kranzberg and Carroll Pursell, Jr., eds., *Technology in Western Civilization*, 2 vols. (New York and London, 1967), is an outstanding work which also has a very useful bibliography. Additional works valuable for their bibliographies are Brooke Hindle, *Technology in Early America* (Chapel Hill, 1966), and David D. Van

Tassel and Michael G. Hall, eds., *Science and Society in the United States* (Homewood, Illinois, 1966).

A fundamental fact about the engineering profession is that it is fragmented into a number of separate, often overlapping professional societies. There are two guides through this thicket: Ralph S. Bates, *Scientific Societies in the United States* (Cambridge, 1945), and Engineers Joint Council, *Directory of Engineering Societies and Related Organizations* (New York, 1963). "Engineering Societies, Their Multiplicity, Their Relatives, Their Duplication," *Technology Review* (July, 1933), 330–331, 350, is still useful.

Histories of professional development are few. Fortunately one of the best deals with the birth of civil engineering in America, Daniel H. Calhoun, *The American Civil Engineer, Origins and Conflict* (Cambridge, 1960). The author's stress on the role of bureaucracy and the tensions between professional and organizational loyalty made Calhoun's work of particular value for this study. Raymond Harland Merritt, "Engineering and American Culture, 1850–1875" (Ph.D. dissertation, University of Minnesota, 1968), is an attempt to extend the story of professionalization in civil engineering down to 1875. Charles Warren Hunt, *Historical Sketch of the American Society of Civil Engineers* (New York, 1897), and the same author's "The Activities of the American Society of Civil Engineers During the Past Twenty-Five Years," *Trans ASCE*, LXXXII (1918), 1577–615, though uncritical, are still useful.

There is as yet no history of the professional development of mining and metallurgy or of the AIME. A. B. Parsons, ed., *Seventy-five Years of Progress in the Mineral Industry, 1871–1946* (New York, 1947), contains information on the history of the AIME. Thomas A. Rickard, ed., *Rossiter Worthington Raymond, A Memorial* (New York, 1920), includes recollections of the founder and secretary of the AIME by his friends. Two other books by Rickard contain material on the professional development of mining engineering: *Retrospect, An Autobiography* (New York, 1947), and *A History of American Mining* (New York, 1932). Rodman Wilson Paul, *Mining Frontiers of the Far West, 1848–1880* (New York, 1963), is an insightful study of the development of mining technology. Thomas Thornton Read, *The Development of Mineral Industry Education in the United States* (New York, 1941), compiles much data on education.

Monte A. Calvert, *The Mechanical Engineer in America, 1830–1910, Professional Cultures in Conflict* (Baltimore, 1967), deals with the early professional development of mechanical engineering and the history of the ASME. The author includes, among other things, a survey of professional development in other areas of engineering and a brief history of early efforts to unite the profession. Joseph Bruce Sinclair, "Science with Practice; Practice with Science: A History of the Franklin Institute, 1824–1837" (Ph.D. dissertation, Case Institute of Technology, 1966), is an outstanding work which presents some material on early professionalism in engineering. Frederick Remsen Hutton, A History of the American Society of Mechanical Engineers from 1880 to 1915 (New York, 1915), is uncritical but contains much information. The ASME has sponsored a series of biographies and autobiographies which contain information on the development of professionalism in mechanical engineering. Included are John Fritz, The Autobiography of John Fritz (New York, 1912); William F. Durand, Robert Henry Thurston (New York, 1929); and Albert W. Smith, John Edson Sweet (New York, 1925). A biography of Alexander L. Holley is badly needed.

A study of electrical engineering similar to those of Calhoun and Calvert for civil and mechanical engineering is also badly needed. Not only is there no history of professional development, but a work that has been widely used, Michael Pupin, From Immigrant to Inventor (New York, 1930), is misleading. James Brittain, in his article, "The Introduction of the Loading Coil: George A. Campbell and Michael I. Pupin," Technology and Culture, XI (January, 1970), 36–57, has demonstrated Pupin's unreliability in this important case. The best available sources on the professional development of electrical engineering in America are a series of unpublished papers written by Mr. Brittain while a graduate student at Case Western Reserve University. Two surveys of the early history of the AIEE are available: "American Institute of Electrical Engineers, Historical Sketch of Its Organization and Work," Trans AIEE, VIII (1891), 601–608, and Edwin J. Houston, "A Review of the Progress of the American Institute of Electrical Engineers," Trans AIEE, XI (1894), 275–284. The IRE has been much better served than the AIEE from the historical standpoint. Among the several historical articles to appear in its publications, "The Genesis of IRE," Proc IRE, XL (May, 1952), 516–520, and Laurens E. Whittemore, "The Institute of Radio Engineers—Fifty Years of Service," Proc IRE, L (May, 1962), 534–554, are worth noting. Some useful information is contained in works that do not deal directly with professional development. Harold C. Passer, The Electrical Manufacturers, 1875–1900 (Cambridge, 1953), and Matthew Josephson, Edison, A Biography (New York and London, 1959), are especially valuable. Also worthy of note is W. Rupert

Maclaurin, *Invention and Innovation in the Radio Industry* (New York, 1949).

While comparatively little has been written on the history of engineering societies, the wealth awaiting discovery is suggested by two outstanding studies. George V. Thompson, "Intercompany Technical Standardization in the Early American Automobile Industry," *Journal of Economic History*, XIV (Winter, 1954), 1–20, sheds much light on the early history of the SAE and its relationship to the industry it served. Bruce Sinclair, "The Cleveland 'Radicals': Urban Engineers in the Progressive Era, 1901–1917" (seminar paper, Case Institute of Technology, 1965), explores the relationship between the Cleveland Engineering Society and the reform administrations of Thomas L. Johnson and Newton D. Baker in Cleveland.

There are a number of sources on early efforts to achieve engineering unity, including the works by Calvert and Hutton cited above. Frederick Haynes Newell, "An Engineering Council Now Almost Forgotten, *Engineering News-Record*, LXXX (April 25, 1918), 806–808, deals with one of the first attempts. Calvin W. Rice, "Joint Engineering Society Activities in the United States," *Mining and Metallurgy*, II (July, 1921), 8–9, is a convenient summary.

The sparse literature on the history of engineering societies may be supplemented by studies of special aspects of professionalism, such as ethics. Among the few frank discussions of this issue is a pamphlet, Morris L. Cooke, *Professional Ethics and Social Change* (New York, 1946). Clyde L. King, ed., "The Ethics of the Professions and of Business," *The Annals of the American Academy of Political and Social Science*, CI (May, 1922), contains several noteworthy articles, especially the ones by Morris L. Cooke and Frederick Haynes Newell. One of the most controversial ethics cases is discussed in a pamphlet, Bernhard F. Jakobsen, *Ethics and the American Society of Civil Engineers* (Los Angeles, 1955). H. A. Wagner, "Principles of Professional Conduct in Engineering," *The Annals of the American Academy of Political and Social Science*, CCXCVII (January, 1955), 46–58, is a recent survey.

Codes of ethics are but one aspect of the organization and operation of engineering societies. Membership requirements, governmental and committee structure, and other matters are also significant for professional development. John W. Lieb, Jr., "The Organization and Administration of National Engineering Societies," *Trans AIEE*, XXIV (1905), 283–296, stresses membership requirements, which I have taken to be the key to professional orientation. F. R. Hutton, "The Mechanical Engineer and the Function of the Engineering Society," *Proc ASME*, XXIX

(December, 1907), 597–632, is also useful. Data on employer subsidies of engineering societies may be found in E.J.C. Subcommittee on Survey of Employer Practice Regarding Engineering Graduates, "Employer Practice Regarding Engineering Graduates," *Mechanical Engineering*, LXIX (April, 1947), 306–308, and the same committee's "Supplemental Report," *Mechanical Engineering*, LXX (January, 1948), 13–16. Morris L. Cooke, "On the Organization of an Engineering Society," *Mechanical Engineering*, XLIII (May, 1921), 323–325, 356, is insightful if polemical. Some interesting information came out in the discussion of this paper: "Organization of an Engineering Society Discussed at A.S.M.E. Spring Meeting," *Mechanical Engineering*, XLIII (August, 1921), 538–540.

Chapter III is based on approximately twelve hundred expressions of social and professional ideas by engineers, the great bulk of which I found in the engineering press. There is no guide to this literature, but several of the more important addresses are reprinted in Carroll W. Pursell, Jr., ed., *Readings in Technology and American Life* (New York and London, 1969). The presidential addresses before the founder societies provide a good index of engineering thought, preserved conveniently in the annual *Transactions* of these societies. A particularly rich vein is provided by the publications of the ASME in the period from 1908 to about 1940. In interpreting this information I found the works of a number of scholars to be of value, too many to attempt to cite them here.

The story of the professional agitation in the AIEE and the AIME is based on the engineering journals. Some further insights into the development of professional attitudes are provided by books published by electrical and mining engineers. Charles P. Steinmetz, *America and the New Epoch* (New York, 1916); Walter N. Polakov, *The Power Age, Its Quest and Challenge* (New York, 1933); and John Mills, *The Engineer in Society* (New York, 1946), suggest the existence of a reform tradition that found little expression in the official publications of the AIEE. Herbert C. Hoover, *Principles of Mining* (New York, 1909), contains much information on the organization and structure of mining engineering.

We need a biography of Frederick Haynes Newell. A great deal about Newell and the conservationist background of his engineering reforms is contained in an outstanding work, Samuel P. Hays, *Conservation and the Gospel of Efficiency* (Cambridge, 1959). I found two of Hays's major conclusions particularly helpful. He shows that conservation was a form of centralized, scientific planning. He also points out that scientific

conservationists were more concerned with the efficient use of resources than with their ownership. Further information on Newell is
contained in Allen B. McDaniel, "Frederick Haynes Newell," *Trans
ASCE*, XCVIII (1933), 1597–600, and Arthur P. Davis, "Frederick
Haynes Newell," *Dictionary of American Biography*, XIII, 456–457.
Newell's activities can be traced in his pocket diaries, previously mentioned, but for his thoughts one must consult his writings. Fortunately,
a complete list of his publications in these years is contained in a series
of pamphlets published by the University of Illinois under the title
Books and Articles Published by the Corps of Instruction. Newell's bibliographies may be found in the issues for May 1, 1915–April 30, 1916;
May 1, 1916–April 30, 1917; May 1, 1917–April 30, 1918; May 1,
1918–April 30, 1919; and May 1, 1919–April 30, 1920. These lists are
particularly valuable since Newell included his anonymously written
articles in his bibliographies, possibly being unaware of the fact that
they were being published. Newell was often much more outspoken in
his unsigned publications.

The best single source for the discontent within the ASCE is *Engineering News*. Of particular importance are the letters to the editors;
these constituted almost the only channel through which the younger
men could express their opinions. Percy E. Barbour, "Secret Technical
League for Licensing Engineers," *Mining and Metallurgy*, II
(December, 1921), 3–4, is a hostile review of the activities of the Technical League. "Pioneering in Professional Welfare," *Professional Engineer*, XIX (April, 1934), 2–23; (July, 1934), 2–23; (October, 1934), 2–24,
constitutes an official history of the AAE. William G. Rothstein, "The
American Association of Engineers," *Industrial and Labor Relations
Review*, XXII (October, 1968), 48–72, is an excellent study, but came to
my attention too late for this work. Gardner S. Williams, "Engineering
Cooperation Outside the National Societies," *Bulletin of the Federated
American Engineering Societies*, II (April, 1923), 5–6, includes a survey
of the cooperation movement by an engineer hostile to Newell. Bruce
Sinclair, "The Cleveland 'Radicals': Urban Engineers in the Progressive
Era, 1901–1917" (seminar paper, Case Institute of Technology, 1965),
previously mentioned, deals with the most militant of the local engi-
neering societies. The history of the Engineering Council is surveyed in
Alfred D. Flinn, "Engineering Council Accomplishments," *Mining and
Metallurgy*, I (April, 1920), 11, and in a pamphlet, J. Parke Channing,
Philip N. Moore, and Alfred D. Flinn, *Engineering Council, A Brief
History* (New York, 1921).

A good starting place for any study of scientific management is Frank

Barkley Copley, *Frederick W. Taylor, Father of Scientific Management,* 2 vols. (New York, 1923). My work has benefited from the existence of several modern, critical studies of scientific management. Hugh G. J. Aitken, *Taylorism at the Watertown Arsenal* (Cambridge, 1960), though dealing with a rather small incident, is sensitive to the deeper philosophical implications of scientific management. Milton J. Nadworny, *Scientific Management and the Unions, 1900–1932* (Cambridge, 1955), is much broader than its title would suggest. It is a detailed study of the internal evolution of scientific management, with emphasis on a shift left that took place after Taylor's death in 1915. Two of the same author's articles help fill out the story: "The Society for the Promotion of the Science of Management," *Explorations in Entrepreneurial History,* V (May, 1953), 244–247, and "Frederick Taylor and Frank Gilbreth: Competition in Scientific Management," *Business History Review,* XXI (Spring, 1957), 23–24. Samuel Haber, *Efficiency and Uplift, Scientific Management in the Progressive Era, 1890–1920* (Chicago and London, 1964), is an insightful study of the reform implications of scientific management. Horace B. Drury, *Scientific Management* (New York, 1922), is still of value. Leon P. Alford, *Henry Laurence Gantt, Leader in Industry* (New York, 1934), and Edna Yost, *Frank and Lillian Gilbreth* (New Brunswick, 1949), are uncritical studies of Taylor associates but contain valuable information. Henry L. Gantt, *Organizing for Work* (New York, 1919), is the best synthesis of Gantt's ideas. Scudder Klyce, *Universe* (Winchester, 1921), which has introductions by John Dewey and Morris L. Cooke, is the best introduction to Klyce's philosophy. For Taylor's own works, a convenient compilation is Frederick W. Taylor, *Scientific Management, Comprising Shop Management, The Principles of Scientific Management and Testimony Before the Special House Committee* (New York, 1947).

There are two biographies of Morris L. Cooke: Kenneth E. Trombley, *The Life and Times of a Happy Liberal, A Biography of Morris Llewellyn Cooke* (New York, 1954), and Jean Christie, "Morris Llewellyn Cooke: Progressive Engineer (Ph.D. dissertation, Columbia University, 1963). Cooke's own story of his struggles within the ASME is detailed in a pamphlet, Morris L. Cooke, *How About It?* (Philadelphia, 1917). I have previously dealt with the interaction between Cooke and Thorstein Veblen in "Veblen and the Engineers," *American Quarterly,* XIV (Spring, 1962), 64–72. Wallace Clark and others, *Fred J. Miller* (New York, 1941), is thin and uncritical, but it is the best sketch of Cooke's closest associate in the reform of the ASME. Mortimer E. Cooley, *Scien-*

tific Blacksmith (New York, 1929), contains little about the internal politics of the ASME, but it does make clear that Cooley was a moderate progressive, a fact also indicated by Cooley's correspondence with Herbert Hoover in the Hoover Papers.

A critical history of public utility regulation is badly needed. The struggle with the utilities was second only to scientific management as a guiding star of Cooke's career. Cooke's declaration of war against the utilities is contained in a pamphlet, Morris L. Cooke, *Snapping Cords* (Philadelphia, 1915). His subsequent relations are suggested in a series of writings, including *Our Cities Awake, Notes on Municipal Activities and Administration* (New York, 1918), and Morris L. Cooke, ed., *Public Utility Regulation* (New York, 1924). Cooke's activities as director of the Department of Public Works of the city of Philadelphia may be best followed in his official reports: Department of Public Works, Philadelphia, *Plain Talk* (Philadelphia, 1914); *Annual Report of the Department of Public Works of the City of Philadelphia, for the Year Ending December 31, 1915* (Philadelphia, 1916). Also helpful are his numerous editorial comments in *Utilities Magazine*. Forrest McDonald, *Insull* (Chicago, 1962), does not deal directly with Cooke, but contains a wealth of information on the organization and influence of the electrical utility interests.

There is no adequate biography of Herbert Hoover. A fair sample of the existing literature is Eugene Lyons, *Our Unknown Ex-President, A Portrait of Herbert Hoover* (New York, 1948), which includes an extensive bibliography. Herbert Hoover, *The Memoirs of Herbert Hoover*, 2 vols. (New York, 1952), has little on Hoover's relationship with the engineering profession; what it does have on engineering is not always reliable. This study, therefore, is based on the primary materials. Two books by Hoover were useful. Herbert Hoover, *American Individualism* (New York, 1929), is a classic statement of Hoover's individualism but is misleading if taken in isolation from his engineering addresses, which provide insights into the roots of his ideological commitments. Herbert Hoover, *Principles of Mining* (New York, 1909), already mentioned, has material on Hoover's conception of his own profession.

There is little of a historical nature on the FAES or the events within the engineering profession that brought it into being. "How the World's Largest Engineering Society Came into Existence," *Mining and Metallurgy*, I (July, 1920), and "Account of the Organization of the Federated American Engineering Societies," *Mining and Metallurgy*, I (July, 1920), 7–10, are contemporary accounts. "History and Review of

Work of Federated American Engineering Societies," *Bulletin of the Federated American Engineering Societies*, II (February–March, 1923), 1–9, is an official sketch.

The best published accounts of the FAES's major investigations are the official reports: Committee on Elimination of Waste in Industry, *Waste in Industry* (New York, 1921), and Committee on Work Periods in Continuous Industry, *The Twelve-Hour Shift in Industry* (New York, 1922). Morris L. Cooke, *Professional Ethics and Social Change* (New York, 1946), already mentioned, includes a brief review of the reaction which crippled the FAES. Donald C. Swain, *Federal Conservation Policy, 1921–1933* (Berkeley and Los Angeles, 1963), is an outstanding study of changing conservation policies. One of the best examples of the repudiation of progressivism is the mystical conservatism of Michael Pupin presented in his *The New Reformation, From Physical to Spiritual Realities* (New York, 1927), and *Romance of the Machine* (New York, 1930). The continued vitality of the progressive tradition is suggested by Charles Beard, ed., *Toward Civilization* (New York, 1930).

Edna Yost, *Modern American Engineers* (Philadelphia and New York, 1952), while uncritical, includes biographies of a number of leaders of the profession, including Ralph E. Flanders and Arthur E. Morgan. Flanders's changing views can best be followed through his articles in the engineering press, but these should be supplemented by his books, *Taming Our Machines* (New York, 1931), and *Platform for America* (New York, 1936). Arthur E. Morgan was a prolific author. His earlier philosophy is suggested by his *My World* (Yellow Springs, Ohio, 1927), and *A Compendium of Antioch Notes* (Yellow Springs, Ohio, 1930). His later views appear in his *The Small Community, Foundation of Democratic Life* (New York, 1942); *The Long Road* (Washington, D.C., 1936); *Design in Public Business* (Yellow Springs, Ohio, 1939); and *Nowhere Was Somewhere* (Chapel Hill, 1946).

Henry Elsner, Jr., *The Technocrats: Prophets of Automation* (Syracuse, 1967), is the best available survey of technocracy. It may be supplemented by the large contemporary literature on technocracy. Among the works I found helpful were Wayne W. Parrish, *An Outline of Technocracy* (New York, 1933); Harold Loeb, *Life in a Technocracy* (New York, 1933); Howard Scott and others, *Introduction to Technocracy* (New York, 1933); and J. George Frederick, ed., *For and Against Technocracy, A Symposium* (New York, 1933). Technocracy's relationship to engineering progressivism is suggested in Joseph Dorfman, *Thorstein Veblen and His America* (New York, 1934), and Leon

Ardzrooni, "Veblen and Technocracy," *The Living Age*, CCCXLIV (March, 1933), 39–42. Walter Polakov, *The Power Age, Its Quest and Challenge* (New York, 1933), suggests the continued influence of Gantt's ideas.

Labor unions for engineers first became a major issue in the 1930s. M. E. McIver and others, *The Technologist's Stake in the Wagner Act* (Chicago, 1944), was one of the earliest surveys. Herbert Northrup, *Unionization of Professional Engineers and Chemists* (New York, 1946), appeared just when the movement was beginning to gain momentum. Richard E. Walton, *The Impact of the Professional Engineering Union* (Boston, 1961), is an excellent study written when the union movement for engineers was declining.

Paul H. Robbins, "National Society of Professional Engineers," *General Electric Review* (March, 1954), 47–49, includes a brief history of the NSPE. The National Society of Professional Engineers, *A Professional Look at the Engineer in Industry* (Washington, D.C., 1955), is suggestive concerning the NSPE's attitudes and policies. T. A. Marshall, Jr., "Engineers Joint Council," *Mechanical Engineering*, LXXVI (July, 1954), 582–584, surveys the successor to the AEC. The best source I have found on the National Academy of Engineering is Eric A. Walker, "The Work of the National Academy of Engineering, U.S.A.," *The Journal of the Royal Society for the Encouragement of Arts, Manufactures and Commerce*, CXXII (May, 1969), 385–390.

INDEX

Academy of Engineering, proposal for, 45
Adams, Comfort A., 66
Air pollution. *See* Smoke abatement
Alford, Leon P.: and American Engineering Council study (1931), 231; proposes Committee on Aims and Organization, 170; represents American Society of Mechanical Engineers at organizing conference, 188; on scientific management, 140–141; and 12-hour day report, 195; and waste study, 194
Alienation, and scientific management, 136
American Association of Engineers, 109, 239; acts on collective bargaining, 123; adopts licensing principle, 123; competes with Joint Conference Committee, 187; declining influence of, 211; fails to unify engineers, 127; founding of, 122; growth of, 127, 169; and licensing, 236; and local societies, 188; membership decline, 211; opposes society membership in Federated American Engineering Societies, 188–189; opposes union tactics, 123; at organizing conference, 188; reciprocity agreements with local societies, 124; rejects Charles E. Drayer, 210–211; rejects Frederick H. Newell, 210–211
American Electric Railway Association, joins Engineering Council, 127
American Electrochemical Society, 41, 51n
American Engineering Council: abolished, 241; and Air Corps, 230; alliance with business, 209–210; American Society of Civil Engineers joins, 210; approves 5-year program investigations, 208; and business, 207–208, 241; and Committee on Elimination of Waste, 202; condemns technocracy, 228; conservatism of 1930s, 230; and Department of Com-

merce, 209; dogmatism of, 210; establishes "Committee on Engineering and Allied Technical Professions," 241; and federal government, 208–209; founding of, 189; and government engineers, 242; inability to advance engineers' status, 241; inaction of 1930s, 230; and morality, 208; and Muscle Shoals bills, 210; and Muscle Shoals Commission, 209; opposes Boulder Dam, 208–209; opposes Herbert Hoover's policies, 231; and public utility regulation, 210; and railroads, 210; rejects progressive concept of engineers' role, 210; and rural electrification, 210; similarities to Engineering Council, 208; social philosophy of, 208; and Social Security, 241; and standardization, 209; and states' rights, 210; and stream pollution, 210; study consumption, production, and distribution, 230–231; successor organizations, 249; theory of engineers' role, 208; upholds *laissez faire*, 230; voting in, amendments to constitution, 230; and wage regulation, 230. *See also* Federated American Engineering Societies
American industrial ideal, 208
American Institute of Chemical Engineers, 22n, 41, 238
American Institute of Electrical Engineers, 22n, 26, 41, 51n, 61, 70, 71, 82; adopts ethics code, 84–85; adopts nominating committee, 212; board of directors censors minutes, 92; board of directors curbs local sections, 86–87; business control in, 212; businessmen and membership standards, 88; censors publications, 231–232; character of reform faction, 183; on conservation, 85–86; constitution (1901), 80–81; creates Committee on Develop-

271

TA
1